T0231848

CLINICAL BIOCHEMISTRY AND DRUG DEVELOPMENT

From Fundamentals to Output

CLINICAL BIOCHEMISTRY AND DRUG DEVELOPMENT

From Fundamentals to Output

Edited by
Shashank Kumar, PhD

First edition published [2021]

Apple Academic Press Inc.
1265 Goldenrod Circle, NE,
Palm Bay, FL 32905 USA

4164 Lakeshore Road, Burlington,
ON, L7L 1A4 Canada

CRC Press
6000 Broken Sound Parkway NW,
Suite 300, Boca Raton, FL 33487-2742 USA

2 Park Square, Milton Park,
Abingdon, Oxon, OX14 4RN UK

First issued in paperback 2021

Library and Archives Canada Cataloguing in Publication

Title: Clinical biochemistry and drug development : from fundamentals to output / edited by Shashank Kumar, PhD.

Names: Kumar, Shashank, editor.

Description: Includes bibliographical references and index.

Identifiers: Canadiana (print) 20200279599 | Canadiana (ebook) 20200279637 | ISBN 9781771888691 (hardcover) | ISBN 9780367821470 (ebook)

Subjects: LCSH: Clinical biochemistry. | LCSH: Cell culture. | LCSH: Drug development.

Classification: LCC RB112.5 .C65 2020 | DDC 612/.015—dc23

Library of Congress Cataloging-in-Publication Data

Names: Kumar, Shashank, editor.

Title: Clinical biochemistry and drug development : from fundamentals to output / edited by Shashank Kumar, PhD.

Description: Burlington, ON, Canada ; Palm Bay, Florida, USA : Apple Academic Press, [2021] | Includes bibliographical references and index. | Summary: "This book, Clinical Biochemistry and Drug Development: From Fundamentals to Output, focuses on clinical biochemistry fundamentals, cell culture techniques, and drug discovery and development concepts. It deals with three different fields of clinical research: cell culture, clinical biochemistry, and drug discovery and development. The book introduces cell animal and bacterial culture techniques and their potential uses as well as cell culture techniques. The biochemistry aspect of the book covers the principles of clinical biochemistry and biochemical analysis, biochemical aids to clinical diagnosis, measurement, and quality control. The book also presents important concepts in cell membrane receptor signal transduction pathways as drug targets. The drug development focus of the book discusses the fundamentals of human disease and drug discovery. Various in silico, in vitro, and in vivo approaches for drug discovery are examined, along with a discussion on drug delivery carriers and clinical trials. Overall, the volume provides an overview of the journey from clinical fundamentals to clinical output. The book aims to provide an understanding for the students of health science and life sciences. This book involves the fundamentals of clinical biochemistry and use of cell culture techniques. Furthermore the book will discuss the fundamentals of drug discovery and development. Over the entire boo"-- Provided by publisher.

Identifiers: LCCN 2020028875 (print) | LCCN 2020028876 (ebook) | ISBN 9781771888691 (hardcover) | ISBN 9780367821470 (ebook)

Subjects: LCSH: Clinical biochemistry. | Cell culture. | Drug development.

Classification: LCC RB112.5 .C65 2021 (print) | LCC RB112.5 (ebook) | DDC 615.1/9--dc23

LC record available at https://lccn.loc.gov/2020028875

LC ebook record available at https://lccn.loc.gov/2020028876

ISBN: 978-1-77188-869-1 (hbk)
ISBN: 978-1-77463-899-6 (pbk)
ISBN: 978-0-36782-147-0 (ebk)

About the Editor

Shashank Kumar, PhD

Assistant Professor, Department of Biochemistry,
Central University of Punjab, India

Shashank Kumar, PhD, is an Assistant Professor in the Department of Biochemistry at the Central University of Punjab, India. He previously worked as postdoctoral fellow at the Department of Biochemistry at King George's Medical University, India. Dr. Kumar has edited several books, including *Concepts in Cell Signaling; Secondary Metabolites and Functional Food Components; Carbohydrate Metabolism Theory and Practical Approach; Phytochemistry: An In Silico and In Vitro Update*, etc. Dr. Kumar has more than 30 published scientific papers/reviews/editorial articles in various national and international peer-reviewed journals of repute. He has expertise in the areas of free radical biology, cancer biology, cell culture techniques, characterization of plant products, xenobiotic metabolism, and microbiology. He has presented his research findings at more than 25 national/international conferences and has attended many workshops at the most of major universities and medical colleges throughout India. Dr. Kumar is an academic/editorial board member and reviewer of about 25 international journals. Dr. Kumar is a lifetime member of the Italo-Latin American Society of Ethnomedicine and Indian Sciences Congress Association and is member of Publication Integrity and Ethics, London. Dr. Kumar received his BSc, MSc, and PhD degrees in biochemistry from the Department of Biochemistry, University of Allahabad, India.

Contents

Contributors

Imran Ahmad
Environmental Toxicology Group, CSIR-Indian Institute of Toxicology Research, Lucknow–206001, Uttar Pradesh, India

Riaz Ahmad
Biochemical and Clinical Genetics Lab, Section of Genetics, Department of Zoology, Aligarh Muslim University, Aligarh–202002, Uttar Pradesh, India

Swagata Das
School of Basic and Applied Sciences, Department of Biochemistry, Central University of Punjab, Bathinda, Punjab–151001, India

Swastika Dash
School of Basic and Applied Sciences, Department of Biochemistry, Central University of Punjab, Bathinda, Punjab–151001, India

Risha Ganguly
Department of Biochemistry, University of Allahabad, Allahabad–211001, India

Hadiya Husain
Biochemical and Genetics Research Lab, Section of Genetics, Department of Zoology, Faculty of Life Science, Aligarh Muslim University, Aligarh–202002, Uttar Pradesh, India

Surya Kant
Department of Respiratory and Medicine, King George's Medical University, Lucknow, India

Shashank Kumar
School of Basic and Applied Sciences, Department of Biochemistry, Central University of Punjab, Bathinda, Punjab–151001, India

Prem Prakash Kushwaha
School of Basic and Applied Sciences, Department of Biochemistry, Central University of Punjab, Bathinda, Punjab–151001, India

Rebati Malik
School of Basic and Applied Sciences, Department of Biochemistry, Central University of Punjab, Bathinda, Punjab–151001, India

Santosh Kumar Maurya
School of Basic and Applied Sciences, Department of Biochemistry, Central University of Punjab, Bathinda, Punjab–151001, India

Abhay K. Pandey
Department of Biochemistry, University of Allahabad, Allahabad–211001, India

Shiv Govind Rawat
Department of Zoology, Banaras Hindu University, Varanasi–221005, Uttar Pradesh, India

Amit Kumar Sharma
Department of Agriculture, Uttaranchal University, Dehradun, Uttarakhand, India

Atul Kumar Singh
School of Basic and Applied Sciences, Department of Biochemistry, Central University of Punjab, Bathinda, Punjab–151001, India

Sharmistha Singh
Department of Biochemistry, University of Allahabad, Allahabad, India

Abhilasha Tripathi
Department of Respiratory and Medicine, King George's Medical University, Lucknow, India

P. Seshu Vardhan
School of Biotechnology, Jawaharlal Nehru Technological University, Kakinada–500085, Telangana, India

M. S. Sandeep Vedanarayana
School of Biotechnology, Jawaharlal Nehru Technological University, Kakinada–500085, Telangana, India

K. Vishnupriya
Protein Bioinformatics Lab, Department of Biotechnology, Indian Institute of Technology, Madras–600036, Tamil Nadu, India

Mohammad Waseem
Molecular Oncology Lab, Department of Bioscience and Bioengineering, Indian Institute of Technology, Kanpur–208016, Uttar Pradesh, India

Abbreviations

3D	three-dimensional
ACS	American Chemical Society
ACTH	adrenocorticotropic hormone
ADAM	a disintegrin and a metalloprotease
ADH	autosomal dominant hypocalcemia
Ago	Argonaute
AIs	aromatase inhibitors
ALP	alkaline phosphatase
AMACR	α-methylacyl coenzyme a racemase
AP	acid phosphatase
APUD	amine precursor uptake and decarboxylation
ATM	ataxia telangiectasia mutated
ATPase	adenosine triphosphatase
BIOS	biology-oriented synthesis
BMP	bone morphogenetic protein
BNP	brain natriuretic peptide
BPCI	biologics price competition and innovation
BPH	benign prostatic hyperplasia
BSS	balanced salt solution
BUN	blood urea nitrogen
CALI	chromophore-assisted laser inactivation
Cav-1	caveolin-1
CB	cathepsin B
CCDA	Cambridge crystallographic data center
CCL2	C-C motif chemokine ligand 2
CCl_4	carbon tetrachloride
CCR5	chemokine receptor type 5
CD	cathepsin D
cDNA	complementary DNA
CHARMM	chemistry at Harvard macromolecular mechanics
Ci	*Cubitus interruptus*
CL	cathepsin L
CLRW	clinical laboratory reagent water

COI	conflict of interest
COL1A2	collagen type 1 alpha 2 chain
COPD	chronic obstructive pulmonary disease
Cr	creatine
CRP	C-reactive protein
CSF	cerebrospinal fluid
CV	cardiovascular disease
CXCL1	C-X-C motif ligand 1
CXCR3	C-X-C motif chemokine receptor 3
DAB2IP	DAB2 interacting protein
DcR3	decoy receptor 3
DG	diacylglycerol
DHH	desert hedgehog
DOS	diversity-oriented synthesis
DTC	direct-to-consumer
Dvl	disheveled
ECM	extra cellular matrix
ECN	extracellular notch
ECRs	enzyme coupled receptors
EGCs	embryonic germ cells
EGF	epidermal growth factor
EGFR	epidermal growth factor receptor
EGTM	European group on tumor markers
EMT	epithelial mesenchymal transition
EPO	erythropoietin
ER	endoplasmic reticulum
ER	Ephrin receptor
ESCs	embryonic stem cells
FACS	fluorescence activated cell sorting
FBPA	fibronectin-binding protein A
FBPB	fibronectin-binding protein B
FBS	fetal bovine serum
FDA	Food and Drug Administration
FEVR	familial exudative vitreoretinopathy
FGF	growth factor of fibroblast
FGFR	fibroblast growth factor receptor
FMTC	familial medullary carcinoma of thyroid
FOS	function-oriented synthesis

FRAP	fluorescence recovery/redistribution after photobleaching
FRET	fluorescence resonance energy transfer
Fz	frizzled
GABA	g-aminobutyric acid
GAP	GTPase accelerator protein
GDP	heterotrimeric G-protein
GEF	guanosine exchange factor
GFR	glomerular filtration rate
GnRH	gonadotrophin releasing hormone
GPCR	globular protein-coupled receptors
GPLR	G protein-linked receptors
G_q	G-protein
GROMACS	Groningen machine for chemical simulations
GSK-3ß	glycogen synthase kinase-3 ß
GSTP1	glutathione s-transferase P1
HBV	hepatitis B virus
HCV	hepatitis C virus
HDAC	histone deacetylase
HER-2/neu	human epidermal growth factor receptor 2/neu
hESC	human embryonic stem cell
HFE	high iron Fe
HH	hedgehog
HPH	high-pressure homogenization
HSC	hepatic stellate cells
ICLRs	ion channel linked receptors
ICMJE	International Council of Medical Journal Editors
ICN	intracellular domain of notch
ICS	inhaled corticosteroids
IFCC	International Federation of Clinical Chemistry
IGFR	insulin-like growth factor receptor
IGRAs	IFN-ϒ release assays
IHH	Indian hedgehog
InChI	International Chemical Identifier
IND	investigational new drug
iPSCs	induced pluripotent stem cells
IUPAC	International Union of Pure and Applied Chemistry
IVF	*in vitro* fertilization
JAKS	Janus kinase

KSR	knock-out serum replacement
LAM	lipoarabinomannan
LD	lactate dehydrogenase
LDC	lipid drug conjugates
LiCl	lithium chloride
LOINC	logical observation identifier names and codes
Lrp	LDL receptor-associated protein
mAb	monoclonal antibody
MAML	mastermind-like
MBL	molecular bacterial load
MBP	myelin basic protein
MCCs	multilayered cell culture
MCP	monocyte chemotactic protein
MD-2	myeloid differentiation-2
MDR TB	multidrug-resistant tuberculosis
MDR	multidrug-resistant
MEFs	mouse embryo fibroblasts
miRNA	microRNA
MPS	mononuclear phagocyte system
MRI	magnetic resonance imaging
MS	mass spectrometry
MTB	mycobacterium tuberculosis
MTD	maximum tolerated dose
NACB	National Academy of Clinical Biochemistry
NADPH	nicotinamide adenine dinucleotide phosphate
NAFLD	non-alcoholic fatty liver disease
NASH	non-alcoholic steatohepatitis
NCBI	National Center for Biotech Information
NCR	notch3 include cytokine response
NDA	new drug application
NDI	nephrogenic or renal diabetes insipidus
NFκB	nuclear factor kappa-light-chain-enhancer of activated B
NGFR	nerve growth factor receptor
NICD	notch intracellular domain
NLC	nanostructured lipid carriers
NLS	nuclear localization signals
NOX	NADPH oxidase
NR	notch receptor

NSE	neuron-specific enolase
oncomiR	oncogenic miRNA
PAP	prostatic acid phosphatase
PAS	para-aminosalicyclic acid
PAZ	pyrazinamide
PBMCs	peripheral blood mononuclear cells
PC	partition chromatography
PCa	prostate cancer
PCA3	prostate cancer antigen 3
PDA	protein design automation
PDB	protein data bank
PDGF	platelet derived growth factor
PDGFR	platelet-derived growth factor receptor
PEST	proline, glutamic acid, serine, and threonine
PHS Act	public health service act
PI	phosphatidylinositol
PIP2	phosphatidylinositol-4,5-bisphosphate
PKB	protein kinase B
PLD	protein ligand database
PPARγ	peroxisome proliferator-activated receptor γ
pri-miRNA	primary-miRNA
PSA	prostate-specific antigen
PSMA	prostate specific membrane antigen
Ptc	protein comprised
PTCH	protein commonly known as patched
RA	retinoic acid
RANTES	regulated on activation, normal T-cell expressed and secreted
RAS-MAPK	rat sarcoma-mitogen activated protein kinase
RDA	recommended dietary allowances
RER	rough endoplasmic reticulum
RHR	Rel homology region
RISC	RNA-induced silencing complex
RNAi	RNA interference
RNF43	ring finger 43
ROS	reactive oxygen species
RP	retinitis pigmentosa
RSV	respiratory syncytial viral infections
SAHA	suberoylanilide hydroxamic corrosive

SCLC	small cell lung cancer
SD	standard deviation
SERMs	specific estrogen receptor modulators
SHH	sonic hedgehog
SI	International system of units
SLI	suppressor of lineage
SLN	solid lipid nanoparticles
Smad	small worm mothers against decapentaplegic
SMILES	simplified molecular input line entry system
SMO	seven membranes panning smoothened
SOCS	suppressor of cytokine signaling
STATS	signal transducer and activator of transcription proteins
TAD	domain for transcriptional activation
T-ALL	T-cell acute lymphoblastic leukemia
TAT	turnaround time
TB	tuberculosis
Tg	thyroglobulin
TGF-ß	transforming growth factor ß
TGFβ1	transforming growth factor β1
Th1	T helper 1
TIMPs	tissue inhibitors of metalloproteinases
TKB	tyrosine-kinase binding
TLR9	toll-like receptor 9
TNF	tumor necrosis factor
TPA	tetradecanoyl phorbol acetate
TPO	thrombopoietin
TRAIL	TNF-related-apoptosis-inducing ligand
TST	tuberculin skin test
VEGF	vascular endothelial growth factor
VEGFR	vascular endothelial cell growth factor receptor
VMD	visual molecular dynamics
XDR TB	extensively drug-resistant tuberculosis
ZNRF3	zinc, and ring finger 3
ZPA	zone of polarizing activity

Preface

This book *Clinical Biochemistry and Drug Development: From Fundamentals to Output* is exactly the need of the hour in which all the topics are presented in a simple, concise, and coherent manner covering important aspects of clinical biochemistry fundamentals, cell culture techniques, and drug discovery and development concepts. The whole book deals with the three different fields of the clinical research. These fields are comprised of cell culture, clinical biochemistry, and drug discovery and development respectively.

The introduction discusses cell animal and bacterial culture techniques and their potential use as well as introduces bout cell culture techniques. The biochemistry section includes the principles of clinical biochemistry and biochemical analysis, biochemical aids to clinical diagnosis, measurement, and quality control. In addition, the book also deals with concepts in the cell membrane receptor signal transduction pathway as drug targets.

The section on drug development discusses the fundamentals of human disease and drug discovery. Various *in silico, in vitro,* and *in vivo* approaches for drug discovery are discussed. The book includes discussion on drug delivery carriers and clinical trials.

The book presents a different angle of understanding for students of health science and life sciences. This book involves the fundamentals of clinical biochemistry and the use of cell culture techniques. Furthermore, the book discusses the fundamentals of drug discovery and development.

CHAPTER 1

An Introduction to Cell Culture

AMIT KUMAR SHARMA,[1] ATUL KUMAR SINGH,[2] and
SHASHANK KUMAR[2]

[1]Department of Agriculture, Uttaranchal University, Dehradun,
Uttarakhand, India, Tel.: +91 9793857582,
E-mail: amitbiochembhu@gmail.com

[2]School of Basic and Applied Sciences, Department of Biochemistry,
Central University of Punjab, Bathinda, Punjab–151001, India

ABSTRACT

Cells, a fundamental unit of life, are an extremely valuable tool in cellular biology research. Cell culture comprises isolation of cells from an animal or plant and their subsequent growth in a favorable synthetic environment which supports their growth and provides essential nutrients. They help as a vital model system for studying physiological phenomenon and testing of biologically important compounds for use in medical treatments.

1.1 INTRODUCTION

Cell culture is a method of cell line cultivation. In this process, the cells are taken from animal tissue. The cells directly cultivated from animal tissues are called as a primary culture. Cell culture involves the natural environment (*in vivo*) to further growth of a cell in the controlled artificial condition (*in vitro*). Nowadays various cell lines from various animals like humans, rats, and some other mammals, are cultured in the laboratory. Normally animal cell culture and bacterial culture are the same but some characteristic differences also exist between them. The animal cell cultures are sensitive to damage, more delicate, have a lower growth rate and require complex culture media along with some other supplement (Orlandelli et al., 2012).

Animal tissue culture cells are used for diagnostic and research purposes of different diseases especially concerned with the viral infections. Various types of human pathogens including viruses can be detected and identified. Experiments of Harrison, Carrel, and Lindbergh introduce HeLa cell culture. Selection of culture media is an important step in the cell culture. Culture media provides a favorable condition for growth, adhesion, proliferation, and cell survival. A list of scientists contributed in the field of cell culture technology and their contribution has been listed in Table 1.1.

TABLE 1.1   Contribution in Cell Culture Technology

S. No.	Year	Scientist	Contribution	References
1.	1887	Leo Loeb	Achieves cultured cells inside and outside of the body using "tissue culture in body" technique.	
2.	1907	Ross Harrison	Cultured frog embryo nerve cells and monitored the development of fibers.	Harrison, 1910
3.	1910	Montrose Burrows	Cultured adult and embryonic tissue of dog	
4.	1912	Alexis Carrel	Cultured connective tissue cells and showed contractility in heart muscle tissue	Carrel, 1912
5.	1913	Steinhardt, Israeli, and Lambert	Cultured vaccinia virus in the corneal tissue of guinea pig	
6.	1916	Rous and Jone	Used trypsin to detach cells in culture disk	
7.	1946; 1947	White, Fischer	Designed and developed a synthetic media for plant and animal cell culture	White, 1946
8.	1948	Katherine Sanford et al.	First time cloned the L-cells	Sanford et al., 1948
9.	1949	Enders, Weller, and Robbins	Grown Poliovirus in human embryonic cells in culture	
10.	1952	George Gey	Established cancer cell line from HeLa cells	Gey, 1952
11.	1954	Abercrombie and Heaysman	Established contact inhibition between fibroblasts	Abercrombie and Heaysman, 1954

TABLE 1.1 *(Continued)*

S. No.	Year	Scientist	Contribution	References
12.	1955	Harry Eagle	Defined media and described attachment factors and feeder layers	Hayflick and Moorhead, 1961
13.	1956	Little Field	Used hypoxanthin, aminopterin, thymidine medium for selection of cell selection	
14.	1961	Hayflick and Moorhead	Established finite lifespan of normal human diploid cells	Eagle, 1959
15.	1962	Buonassisi	Developed method for the maintenance of tumor cell differentiation	Hayflick and Moorhead, 1961
16.	1968	David Yaffe	Worked on differentiation of normal myoblasts	Buonassisi et al., 1962
17.	1969	Augusti and Sato	Established neuroblastoma cells	
18.	1973	Graham FL and van der Eb AJ	Introduced DNA in mammalian cell culture	Yaffe, 1968
19.	1975	Georges Kohler and Cesar Milstein	Developed first hybridomal cell secreting monoclonal antibody	
20.	1977	Genetech	Developed the first human recombinant protein somatostatin	
21.	1985	Collen	Produced Bacterial recombinant growth hormone accepted for therapeutic use	
22.	1997	Wilmut, Schnieke, and colleagues	Produced transgenic sheep by using mammalian cell culture and nuclear transfer technique	Wilmut et al., 1997
23.	2002	Clonei	Claimed the production of EVE (cloned human baby)	
24.	2009	Nathalie Cartier-Lacave	Combined blood stem cell therapy with the gene therapy.	
25.	2010	Freshney	Developed media for animal cell culture	Freshney, 1987
26.	2012	Maria Blasco	Established the first gene therapy against an aging	
27.	2013	Brysek and co-workers	Developed media for animal cell culture	Bryzek et al., 2013

Isolated cell from the animal tissue is propagated and maintained outside the body in an appropriate artificial environment. Various animal cells can grow outside the body in a medium containing essential nutrient and growth factor. *In vitro* culture of cell, lines do not completely mimic *in vivo* conditions. Proper CO_2, O_2, osmolality, and nutrition should be provided for the culture. Culture conditions should be sterile. Animal cells are cultured in natural or artificial media as per the need of the experiment. Media is the most essential and important requirement of cell culture.

Each cell requires a specific type of media for growth and survival. These characters of the cells are used for various research experiments. Availability of serum-free media and serum- containing media largely promote the cell culture practices. Cells of different animals and organ origin are now growing in various laboratories of the world using. The major purpose of this routine cell culture is to find out requirements for growth, cell cycle, etc. about desired cells. Homogenous culture obtained from primary cell culture is a useful tool to study the origin of the cells.

KEYWORDS

- **adhesion**
- **animal tissue culture cells**
- **cell line cultivation**
- **HeLa cells**
- **pathogens**
- **proliferation**

REFERENCES

Abercrombie, M., & Heaysman, J. E., (1954). Observations on the social behavior of cells in tissue culture: II. "monolayering" of fibroblasts. *Exp. Cell Res.*, *6*(2), 293–306.

Bryzek, A., Czekaj, P., Plewka, D., Tomsia, M., Komarska, H., Lesiak, M., Sieroń, A. L., Sikora, J., & Kopaczka, K., (2013). Expression and co-expression of surface markers of pluripotency on human amniotic cells cultured in different growth media. *Ginekologia Polska*, *84*(12), 1012–1024.

Buonassisi, V., Sato, G., & Cohen, A. I., (1962). Hormone-producing cultures of adrenal and pituitary tumor origin. *Proc. Natl. Acad. Sci., USA, 48*(7), 1184–1190.

Carrel, A., (1912). On the permanent life of tissues outside of the organism. *J. Exp. Med., 15*(5), 516–528.

Eagle, H., (1959). Amino acid metabolism in mammalian cell cultures. *Science, 130*(3373), 432–437.

Freshney, R. I., (1987). *Training Programs*. John Wiley & Sons, Inc.

Gey, G., (1952). Tissue culture studies of the proliferative capacity of cervical carcinoma and normal epithelium. *Cancer Res., 12*, 264–265.

Harrison, R. G., (1910). The outgrowth of the nerve fiber as a mode of protoplasmic movement. *J. Exp. Zool., 9*(4), 787–846.

Hayflick, L., & Moorhead, P. S., (1961). The serial cultivation of human diploid cell strains. *Exp. Cell Res., 25*(3), 585–621.

Orlandelli, R. C., Specian, V., Felber, A. C., & Pamphile, J. A., (2012). Enzymes of industrial interest: production by fungi and applications. *SaBios–J. Health Biol., 7*(3).

Sanford, K. K., Earle, W. R., & Likely, G. D., (1948). The growth *in vitro* of single isolated tissue cells. *J. Natl. Cancer Inst., 9*(3), 229–246.

White, P. R., (1946). Cultivation of animal tissues *in vitro* in nutrients of precisely known constitution. *Growth, 10*, 231–289.

Wilmut, I., Schnieke, A. E., McWhir, J., Kind, A. J., & Campbell, K. H., (1997). Viable offspring derived from fetal and adult mammalian cells. *Nature, 385*(6619), 810–813.

Yaffe, D., (1968). Retention of differentiation potentialities during prolonged cultivation of myogenic cells. *Proc. Natl. Acad. Sci., USA, 61*(2), 477–483.

CHAPTER 2

Animal Cell Culture

AMIT KUMAR SHARMA,[1] ATUL KUMAR SINGH,[2]
MOHAMMAD WASEEM,[3] and SHASHANK KUMAR[2]

[1]Department of Agriculture, Uttaranchal University, Dehradun,
Uttarakhand, India, Tel.: +91 9793857582,
E-mail: amitbiochembhu@gmail.com

[2]School of Basic and Applied Sciences, Department of Biochemistry,
Central University of Punjab, Bathinda, Punjab–151001, India

[3]Molecular Oncology Lab, Department of Biological Sciences, and
Bioengineering, Indian Institute of Technology, Kanpur–226003,
Uttar Pradesh, India

ABSTRACT

Cell culture generally means to grow cells in a precise condition. Cells can be preserved *in vitro* instead of inside the body. This method is relatively easy compared to organs and tissue cultures. For culturing, cells are isolated from the animal system and grown subsequently in a favorable environment which supports their growth and provides all the essential nutrients required for their growth. Cells can be isolated from the tissue or by mechanical cells. They can also be acquired from earlier created cell strains or cell lines. Cell cultures are utilized for numerous purposes such as model system for studying the various biological phenomenon, toxicity testing, cancer research, genetic engineering, and drug sensitivity testing.

2.1 INTRODUCTION

2.1.1 *ASEPTIC METHODS AND GOOD CELL CULTURE PRACTICE*

2.1.1.1 *BASIC ASEPTIC METHODS IN ANIMAL CELL CULTURE*

To culture the animal cells in the laboratory it is necessary that all processes are carried out in aseptic or sterile conditions. It is essential to indicate here that laminar flow facility or sterile rooms offer proper environment. In a nonsterile environment, reliance on aseptic techniques is very high. The basic aseptic method is to confirm that the workplace is clean, wiped down with ethanol and all the equipment used has been sterilized.

If you are working on the bench, use a Bunsen flame to heat the surrounding Bunsen air. This causes the passage of air and contaminants upwards and lowers the chance of contamination in open vessels. Open all the bottles and do all exercises in this area only, swab the top and neck of bottles with ethanol before opening, flame all bottleneck and pipettes through passing them quickly in the hottest part of flame. This action is not needed with sterile, wrapped, pipettes, and plastic flasks, do no place caps and pipettes on the bench; exercise to the bottle tops with the little finger while holding bottles during pouring or pipetting, work either left to right or conversely, so that every material to be used is on one side and, once finished, is kept on the other side of the Bunsen burner. Manipulate flasks and bottles carefully. The tops of bottle and flask should not be contacted by the operator. Touching of open vessels should also be prevented during pouring. If necessary practice pouring from one container to another at a distance of 5 mm between the two vessels, clear up spills immediately and always keep the work area clean and tidy. Dispose the glassware in appropriate bins and discard used plastic ware in marked polythene bags for autoclave or incineration. All glassware or plastic- ware used for infectious work must always be autoclaved before incineration. Re-usable glassware should be immersed in disinfectant whilst waiting to be autoclaved.

2.1.1.2 *ANIMAL CELLS: BASIC CONCEPTS*

Another underestimated problem in cell culture is the use of antibiotics. Continuous utilization of these compounds is not advised as it favors the growth of resistant microbial strains that become difficult to eradicate. It develops the requirement for potent antibiotics which could be more toxic for animal cells. Furthermore, the utilization of these compounds tends to mask contamination at reduced levels. When contamination occurs, it is recommended to discard the culture and start working with contaminant-free stock. If not possible, antibiotics could be applied to eradicate the contamination. Nevertheless, viral contaminations cannot be treated since they do not respond to antibiotics. Removal of viral contamination by centrifugation or other separation techniques cannot be done since they are intracellular parasite. If virus-free stock does not exist, a risk evaluation should be performed before continuing the work with the infected cell line.

A cell culture research laboratory, in which activities are limited to manipulation of established or pathogen-free derived cell lines, is a relatively safe workplace. Major risks are related to potential injuries resulting from liquid nitrogen or glassware accidents. Activities involving routine pathogen carrying animal cell lines or primary culture derived from infected animal's alarms potential risk of infection to the operator. Viruses present the highest contamination risk, but many bacteria, fungi, mycoplasmas, and parasites are also be harmful for operator. Continuous cell lines not derived from humans or primates and well-characterized diploid human cells lines with a finite lifespan (for example, MRC-5 cells) are regarded as low-risk cell lines. Poorly categorized mammalian cell lines are regarded as medium risk cells. Human and primate tissue cells, cell lines carrying endogenous pathogens, and cell lines manipulated after experimental infections are treated as high-risk cell lines.

2.1.1.3 CHECKING AND PREVENTION FOR CONTAMINATION

Perhaps the biggest pre-occupation of cell culture biologists is to prevent contaminations in tissue culture. The media used to grow cells offers exceptional nutrition for unwanted organisms also. Contamination of cell culture with bacteria, fungi, and mycoplasma often results in a great loss of time and money. An additional problem is the cross-contamination of cells with other cell lines. Before we inspect how to keep contamination at the lowest, we need to consider briefly how we would detect the contamination. Regular checking of aliquots from culture flasks under a microscope can reveal some type of contamination, for example, yeast, and fungi can be seen easily. Bacteria can be Gram stained or plated out on blood agar plates. The general cloudiness of the medium also indicates yeast or bacterial contamination. Mycoplasma contamination is more insidious as the medium of the culture flask will not be cloudy. These organisms are not visible under an ordinary microscope and most mycoplasma species are very difficult to grow on agar plates. Examination of mycoplasma is time-consuming but needs to be performed on a consistent basis because cross- contamination is very easy. DNA stains or molecular probes can be applied to examine their presence in cells, the contaminated cells need to be discarded.

Most of the contamination arises from culture media and its constituents or from inadequate cleaned glassware. Routine sterility checks on medium, serum, and nutrients are recommended. To perform this, a small proportion of medium needs to be incubated at 37°C and another similar aliquot simultaneously incubated at room temperature for one week before the utilization of medium. If these small samples are contaminated, the medium needs to be autoclaved and discarded. Application of antibiotics in the medium assists to sustain the problem up to some level but this cannot replace good aseptic technique and careful monitoring. Generally, if contamination occurs, affected cultures need to be discarded in 2.5% hypochlorite solution. Media bottles which are known, or suspected for contamination should be discarded as well. Some basic preventive measures can be applied for sterilizing procedure (autoclave and oven procedures), sterility of laminar flow hood, consistent checks on cultures; each worker should use their own set of media, and discarder for contaminated cultures instead of trying to decontaminate them.

2.1.1.4   *PREPARATION OF PRIMARY CULTURES*

Every time it is not necessary to disaggregate tissue before culture. Some embryonic tissue can be cultured by keeping whole tissue on the surface of flask and different cells will proliferate from the whole tissue. Some days later, replenish the medium and remove the original tissue to permit the continuous growth of cells. The technique is permissible for some tissues only but it is not appropriate to grow specific cell from a tissue.

2.1.1.5 *TISSUES THAT DO NOT NEED ENZYMATIC DISAGGREGATION*

Primary explant tissues can be treated as follows:

- Excess blood is cleaned by washing with a sterilized balanced salt solution (BSS). If the tissue needs to be transported it can be stored in BSS or medium.
- Undesirable material like cartilage and fat is cut off and rest is finely chopped by pair of crossed scalpels.
- The cell suspension is shifted to a sterilized 50 ml centrifuge tube along with the buffered saline and permitted to settle down.
- The medium is carefully pipetted off and the pellet is washed in fresh BSS. The cells are permitted to settle down again or the suspension is centrifuged at 1,000 rpm for 5 min.
- The cell pellet is again suspended in 10–15 ml of medium and the suspension is aliquoted into 2 or 3 of 25 cm^2 flask. If the parts of tissues are still large, the suspension can be passed through a sieve before culture. The culture is incubated for 18–24 hours at suitable temperature.
- If the parts of tissue are adhered, the medium needs to be changed weekly until a significant outgrowth of cells has occurred.
- The original tissue pieces which showing outgrowth should be picked off and shifted to a fresh flask.
- The medium in the previous flask is replaced and the cells are cultivated until they cover at least 50% of the surface. They can then be subcultured if required.

All steps involved should lower the chances of tissue, media, and vessel contamination. Thus, all centrifuge, tubes dissection tools, and reagents should be sterilized before use. This approach is used for small parts of tissue such as skin biopsies. Fibroblasts, glial cells, epithelium, and myoblasts migrate out of the tissue very successfully. Cells should be selected as some cells grow rapidly and some are adhered. Tissue required enzymatic disaggregation is kept whole or in small pieces as mentioned above.

Though tightly packed tissue is difficult to disaggregate and lengthy contacts with digestive enzymes like trypsin. It generally causes loss of the living cells. Large tissue needs to be divided into smaller pieces for primary explant and digested with trypsin for 30 min at 37°C. The trypsin is neutralized by the application of serum. The quantity of trypsin required to attain the satisfactory release of cells is tissue-dependent and also on the activity of enzyme preparation. Trypsin suppliers often provide useful guidance for the application of preparations. Typically, 0.25% w/v trypsin solution is satisfactory.

By adding a lot of protein (in the form of serum), the enzyme begins to hydrolyze these proteins rather than the proteins which bind cells together. What we are really doing is 'diverting' the enzyme to another target so that it cannot attack the extracellular matrix. After trypsinization, a medium comprising serum should be added to neutralize enzyme activity. If trypsin is applied to disaggregate the whole tissue, the dissociated cells need to be harvested after about 30 minutes and washed to remove trypsin. The remained tissue might be trypsinized. Another approach is to soak overnight the piece of tissue in cold trypsin at 4°C. The cold trypsin infiltrates throughout the tissue but have minimal action at this temperature. The tissue can be disaggregated by increasing the temperature up to 37°C for 30–40 min, and enzyme activity is nullified by adding a medium with serum as before.

Collagenase and versene (phosphate-buffered saline + EDTA) are also used to disaggregate tissue. Both are gentle in action in comparison to trypsin but costly due to the requirement of a large amount. Collagenase or versene can also be applied together with trypsin. A number of other enzymes (for example pronase and dispase) have also been applied to disaggregate tissues in some laboratories. After disaggregation, the cells are washed in fresh BSS/medium and seed in 24 cm² flask. The cell layer can be supplemented with a fresh medium after 48 hours. This culture is called as a primary culture.

2.1.1.6 *REMOVING NON-VIABLE CELLS FROM THE PRIMARY CULTURE*

In an anchorage-dependent or adherent primary cell culture, the non-viable (dead) cells can be detached by removing the medium and washing the cells with buffer before addition of fresh medium. Non-viable cells will not able to adhere to the surface. If the cells need to be cultured in liquid medium, the non-viable cells become diluted as the viable cells duplicate. If it is essential to eliminate the non-viable cells from the liquid medium, the cells can be layered on Ficoll or 'lymphoprep' and centrifuged at 2,000 rpm for 15–20 min. The non-viable cells will descend into the bottom and viable cells can be recovered from the medium-Ficoll interface

2.1.1.7 MAINTAINING THE CULTURE

If a primary culture does not have immediate use, it can be sub-cultured to establish a cell line. Cell lines may be of a very limited lifespan or they may divide few times before they become senescent. Some cells, like neurons and macrophages, do not proliferate *in vitro* and they can only be utilized as primary culture.

2.1.1.7.1 Sub-Culturing From Primary to Secondary Cell Culture

A primary culture comprises a very diverse population of cells from the novel explant. Some of the cells will die, some are unable to grow, and others will grow rapidly and become the leading cell type. In a sub-culture derived from primary cell culture, the leading cell types will emerge as even more dominant. Sub-culture allows us to create more homogeneous cell population.

The culture produced after the sub-culture of a primary culture is referred as secondary culture. Sub-culturing permits us to create cell lines from the original explant. These cell lines can be again sub-cultured, categorized, and cloned. Creating a cell line has some obvious advantages. A uniform population of categorized cell could be grown on large scale and produce uniform replicates. Few cell lines may lose differentiated features while growing *in vivo* and are prone to genetic instability, mostly if they divide rapidly.

2.1.1.7.2 Producing Cell Lines of a Specific Cell Type

There are two general considerations during establishment cell lines of a specific cell type. First, of course, we need to select the appropriate tissue of primary explant stage. Second, not every cell will grow uniformly well in the identical medium. Thus in principle, during the selection of medium we should carefully provide most suited condition for cell of interest.

2.1.1.7.3 Propagating a Cell Line

After the cell line establishment, it requires to propagate to produce enough cells for characterization, storage, and experiment. A cell line is provided a name or code that reflects its origin (for example HuT, Human T-cells) and, in cases where more than one line was produced from the similar origin, a cell line number is also provided (for example HuT 78). If cells from this line are being cloned, then a clone number needs to be provided, e.g., HuT 78 clones 6D5.

In cases where the cell line is most probably to be viable for only a certain number of sub- cultures or generation each generation should be noted, for example, HuT 78 clone 6D5/2 or HuT 78 clone 6D5/3. If the culture becomes confluent (i.e., the cells acquiring 60–70% area of the growth surface) it needs to be shifted into a holding or maintenance medium. It offers sufficient nutrients to keep the cells viable and healthy but lowers the rate of replication to check the overgrowth of cells. The general method is to utilize the same basic medium with the reduced the quantity of serum. If cells are needed for experiments or storage, the amount in the flask can be split or divided into 2–4 new flasks dependent on the vigor of cell growth. The amount of cells cultured in suspension is very easy to split. This is achieved as follows:

- Stand the flask upright for 1–5 min to permit cells to settle.
- Aseptically remove as much medium as possible without upsetting the cells. A 10 ml graduated pipette is best. Discard used medium into a waste pot for autoclave.
- Re-suspend cells in the residual volume, measure, and divide between the required numbers of flasks.
- Top up each flask with fresh growth medium and incubate as normal.

Record the cell line code, clone number, and the passage number. Cells that proliferate as monolayer adhering to the flask surface are a little more difficult to split. The steps are as follows:

- Remove medium by pouring it off aseptically.
- Rinse the monolayer with previously warmed (37°C) PBS or BSS comprising 1 mM EDTA; remove the medium off into the discard pot.
- Add 10 ml of 0.25% (w/v) of trypsin solution in saline. Tilt the flask so that all the cells become covered with trypsin.
- Remove the trypsin and incubate the flask at 37°C for 1–10 min. Cells that are very sensitive to trypsin needs to be examined after 1 min.
- Once the cells start to round up and are sliding off the flask surface the trypsin is neutralized by adding 5–10 ml of serum-containing medium.

- Wash all the cells down from the sides of the flask and aspirate gently using a pipette. This should break up aggregates and allow easier cell counting if required. Avoid creating froth as this can result in contamination and cell damage.
- A small aliquot (0.2 ml) of the cell suspension can be used for counting if required. Otherwise, measure the total volume of cell suspension and divide between the required numbers of flask and top up with fresh medium and incubated as before.
- Remember to record the cell line code, clone number and how many times it has been divided (the passage number).

The interval between sub-culturing and exchanging the medium will depend on cell line and its growth rate. Rapidly dividing cell lines like HeLa and VERO needs to subcultured at least once a week with a medium change in between. Fibroblast HEL cells can be sub-cultured once every 10–14 days. The growth level of fibroblasts can be improved by the adding serum in the medium.

2.1.1.8   *QUANTITATION OF CELLS IN CELL CULTURE*

For proper experiments, it might be essential to calculate the cell numbers before, after, and during the experiment. Day to day preservation of cell lines also requires a quantitative valuation of cell growth so that optimal cell density for sub-culturing and storing of cells can be decided.

We can categorize the systems existing for cell growth determination into two sub-groups.

These include direct and indirect methods.

2.1.1.8.1   *Direct Approach*

In the direct approach, cell numbers are calculated directly either by counting via a counting chamber or with an electronic particle counters.

2.1.1.8.2   *Indirect Approaches*

In the indirect approaches, properties like DNA or protein content linked to cell numbers are utilized to estimate biomass.

2.1.1.8.3   Counting Chambers

A hemocytometer (Improved Neubauer) is the simplest and cheapest method of counting. A coverslip marginally wet at the edges is kept on the counting grid such that interference pattern (Newton's rings) is formed. Introduce the medium containing cells under coverslip from a Pasteur pipette by capillary action. The fluid will run into grooves of the chamber to take away the excess. The cells in the central grid area can then be counted. The hemocytometer is designed in a way that when properly set up an area of 1 mm^2 and 0.1 mm deep is filled. This is a central area of 25 small squares bounded by triple parallel lines. Every 25 square is again divided into 16 to aid counting. For routine subculture about 100–300 cells need to be counted, but 500–1,000 is more precise counts. For the better Neubauer chamber, the cell concentration, calculated from the number of cells (n) in the central 25 square areas (1 mm^2, depth 0.1 mm), is n × 10^4 cells/ml. Viability determinations can be easily carried out by diluting cells with dye.

This counting method is open to an error in a number of ways. Errors may be introduced by incorrect preparation of the chamber, incorrect sampling of cells, and aggregation of cells. However, it has the advantage of visualization of the cell morphology.

2.1.1.8.4   Coulter Counters

Electronic particle counters are made up of two electrodes alienated by a small orifice. If a potential is applied to the electrodes, current will pass between them through the buffer in the orifice. The quantity of current will depend on the conductance (dielectric constant) of the buffer. As the particle enters the orifice, the conductance of the solution between electrodes would be reduced and detected electronically. The change in current flow is dependent on the size of particle, the difference in the dielectric constant (conductivity) of the particles and suspending buffer.

Cells in liquid medium are drawn via a fine orifice in the Coulter counter and, as each cell passes through produces a change in the current flowing across the orifice. Each change is recorded as a pulse and are sorted and counted. The size of pulse is proportional to volume of the particle passing through. Thus, the signals of varying size are produced. The pulse height threshold could be set to eliminate electronic noise and weak signals produced by debris. Coulter counter is widely used and it provides rapid results. It requires high cell numbers to give accurate count. Another drawback is that they cannot differentiate between dead, viable, and clumps of cells. It may record as a single pulse leading to inaccurate count.

2.1.1.8.5 Indirect Methods for Determining Cells in Culture

Other methods of quantification like radioisotope labeling and evaluation of total DNA or protein are applied less frequently. They are applicable for cells grown in microwell plates or in hanging drop culture. The cells can be mixed in situ, stained, and counted by eye under a microscope. DNA and protein assays are imprecise, predominantly if cells are multinucleated. They do not differentiate into viable and non-viable cells.

2.1.1.9 CELL VIABILITY DETERMINATION

When cells are recently taken from a tissue or confluent monolayers are sub-cultured, the percentage of living, or viable, cells needs to be determined before use. Most often, it is estimated by determining of membrane permeability under the postulation that a cell with a permeable membrane has suffered severe, irreversible damage.

2.1.1.9.1 Trypan Blue Exclusion

This is a quick test for gross damage conveniently combined with cell count. Following is the procedure step by step:

- Mix a small proportion of cell suspension with an equivalent volume of 0.4% trypan blue solution.
- Within 1–5 min, introduce suspension into a hemocytometer chamber.
- Non-viable cells will appear blue, with the nucleus dark stained.
- Count the living (unstained) cells and the entire number of cells.
- Express the number of living cells as a fraction of the total. Analyze the cell number after multiplying by 2 to allow dilution with the dye.

2.1.1.10   SAFE LABORATORY PRACTICES

The subsequent recommendations are given for safe laboratory practices, and they should not to be understood as an entire code of practice. Refer your institution's safety committee and follow local rules and regulations related to laboratory safety.

For more information regarding microbiological practices and for biosafety level guidelines, consult to Biosafety in Microbiological and Biomedical Laboratories, 5th Edition at www.cdc.gov/od/ohs/biosfty/bmbl5/bmbl5toc.htm.

- Constantly wear proper personal protective gear. Change contaminated gloves, dispose used gloves with other contaminated laboratory waste.
- Clean your hands after working with hazardous materials and before leaving the laboratory.
- Do not eat, smoke, drink, handle contact lenses, use cosmetics, or store food for human consumption in the laboratory.
- Follow the institutional policies of the safe handling of needles, scalpels, pipettes, and broken glassware.
- Minimize the formation of aerosols and/or splashes.
- Clean all workbenches before and after your experiments, and instantly after any spill or splash of infectious substances with a disinfectant. Clean laboratory apparatus regularly.
- Decontaminate all infectious substances before disposal.
- Report every incident that might result in the exposure of infectious material to proper personnel (e.g., laboratory supervisor, safety officer)

2.2   TYPES OF ANIMAL CELL, CHARACTERISTICS, AND MAINTENANCE IN CULTURE

2.2.1   TYPES OF ANIMAL CELL

Animal cells are generally defined on the basis of tissue from which they are derived. They have a characteristic shape which can be detected easily with the help of light microscope. These are generally derived from epithelial, connection, blood, and lymph, muscular, and neuronal tissue.

2.2.1.1   EPITHELIAL TISSUE

It comprises a layer of cells that surrounds organs and line cavities such as skin and the lining of alimentary canal. The epithelial cells proliferate well in culture as a single cell monolayer and have typical cobblestone appearance.

2.2.1.2   CONNECTIVE TISSUE

It forms main structural part of animals, containing a bone cartilage and fibrous matrix. The tissue contains fibroblasts, the most widely applied cells in laboratory cell cultures. Fibroblasts are bounded by the fibrous protein collagen in the connective tissue. Initially these cells are spherical, during trypsinization from the tissue but elongate in typical spindle-shape on attachment to a solid surface. Fibroblasts has excellent growth characteristic which makes it a 'favorite' cell for culture establishment. Fibroblast and epithelial cells are comparatively simple to culture and their growth rate is about 18–24 hours.

2.2.1.3   BLOOD AND LYMPH

It contains a variety of cells in suspension. Few of these will remain growing in a culture suspension. These include lymphoblasts the WBCs and are utilized widely in culture due to their capability to release immunomodulating compounds.

2.2.1.4   MUSCLE TISSUE

It comprises a series of tubules created from precursor cells fused to form a multinucleated composite and contain structural proteins like actin and myosin. The precursor of these tissues is myoblast, which has the capability to differentiate into myotubes. This phenomenon can be detected in culture.

2.2.1.5   NERVOUS TISSUE

Nervous tissues are consisting of characteristic shaped neurons (responsible for transmission of electrical impulse) and secondary cells, such as glial cells. Neurons are extremely differentiated and they are unable to divide in culture. Though, the adding nerve growth factor to neuronal culture might cause creation of cytoplasmic outgrowth called neurites. Few characteristics of nerve cells might be observed in neuroblastomas. They are cancer cells that carry out cell growth in culture.

2.2.2 FUNDAMENTALS OF ANIMAL CELL CULTURE: MAINTENANCE IN CULTURE

Cell microenvironment involves animal cell culture having living cells of animals in an isolated environment. The physicochemical properties of microenvironment are capable to match the physiological conditions of cells origin. Few essential features for a culture medium (Freshney, 2015; Bryzek et al., 2013), are described in preceding headings:

2.2.2.1 SUBSTRATE

The mainstream of cultured cells proliferates as a monolayer on an artificial substrate. Although few transformed cell lines and hemato-poietic cells were grown in liquid medium. The substrate needs to be appropriately dispersed to permit cell adhesion and proliferation as well as secretion of cell adhesion molecule. Most used substrates comprise glass and plastic for its optical characteristics and regularity. Although synthetic fiber (applied in creating scaffolds of two or three dimensions), and metals are applied to transfer the sample to electron microscopy (Freshney, 2015).

Additionally, the substrate property could be improved by treating it with extracellular matrix collagen and fibronectin. Sometimes tissue-derived extracellular matrix and polymers such as poly-L Lysine or commercial matrices are also applied (Freshney, 2015; Ross et al., 2012).

2.2.2.2 HYDROGEN POTENTIAL (pH)

Most animal cells grow at optimal pH between 7.0 and 7.4. Nevertheless, this can differ remarkably in the transformed cells (Chaudhry et al., 2009).

2.2.2.3   BUFFERING, CARBON DIOXIDE AND BICARBONATE

CO_2 dispersed in media established equilibrium with HCO_3-ions and decrease the pH (Freshney, 2015). In spite of its poor buffering ability at physiological pH, bicarbonate is generally applied due to its low toxicity and nutritional property in the culture. The pH of culture media is buffered by two circumstances, the opening of boxes, where CO_2 entry increases pH; and CO_2 and acid production because of high cell concentration resulting into decreased pH (Freshney, 2015; Frahm et al., 2002).

2.2.2.4   OXYGEN

Most cells require oxygen for respiration *in vivo*. However, some of the transformed cells are anaerobic. Oxygen is still obligatory and its concentration differs depending on the type of culture. The low concentration is better for majority of the cells (Freshney, 2015).

2.2.2.5   OSMOLARITY

The mainstream of the cultured cells are tolerant to osmotic pressure. Human plasma osmolality is near 290 mOsm/kg, reasonable to assume the optimal level for cells *in vitro,* though it can differ for different species. Osmolarity between 260 mOsm/kg–320 mOsm/kg is suitable for most cells, but once maintained should be kept constant at ±10 mOsm/kg (Freshney, 2015).

2.2.2.6   TEMPERATURE

The optimal temperature for the cells of animal origin and anatomic variation regulates the optimum temperature for cell culture. For example, the skin and testicles, have usually, a lower temperature than the rest of the body cells (Lee et al., 2013). It is essential to consider a few safety factors to sustain the minimum minor error in the regulation of incubator. One needs to have an alarm incubator for 1° above and below the incubation at the preferred temperature. The temperature for most warm- blood animals' cell line is 37°C.

2.2.2.7   VISCOSITY

The viscosity of culture media is affected by serum content. It is significant when a cell suspension is stirred or when cells are detached after trypsinization. Damage in such situations can be minimized by increasing the viscosity of the culture medium (Freshney, 2015).

2.2.2.8   AMINO ACIDS AND VITAMINS

Essential amino acids are a necessity for the cells cultured together with cysteine, arginine, glutamine, and tyrosine. The requirement of amino acid varies from one cell to another type of cell. Eagle's minimum essential media contains water with soluble vitamins (group B, choline, folic acid, inositol, nicotinamide, excluding biotin), and other requirements are derived from serum (Genzel et al., 2004).

2.2.2.9   IONS AND GLUCOSE

Ions found in the cell culture media include Na^+, K^+, Mg^{2+}, Ca^{2+}, Cl^-, SO_4^{2-}, PO_4^{3-} and HCO_3^-. They contribute to the media osmolarity. Glucose serves as an energy source in most of the media (Kwong et al., 2012).

2.2.2.10   ORGANIC SUPPLEMENTS

Media can be supplied with many components, like protein, peptide, nucleoside, citric acid intermediates, pyruvate, and lipids. They usually occur in complex media and reduce serum use. They act as helpers in the cloning and maintenance of specialized cells (Freshney, 2015).

2.2.2.11   HORMONES AND GROWTH FACTORS

They are generally added to serum-free media or provided by the added serum. The main factors found are growth factor derived from platelet, fibroblastic growth factor, epidermal growth factor (EGF), vascular endothelial growth factor (VEGF), and insulin-like growth factor. The most common hormones used in cell culture are insulin and hydrocortisone (Freshney, 2015; Stryer, 1995; Onishi et al., 1999; Aden et al., 2011).

2.2.2.12 *ANTIBIOTICS AND ANTIFUNGAL*

They are applied in conjunction with laminar flow hoods or security flow hoods to reduce the frequency of bacterial and fungal contamination (Freshney, 2015).

2.2.2.13 *SERUM*

Serum present growth factors that promote cell proliferation besides adhesion factors and antitrypsin activity that promote cell adhesion. It is also a source of minerals, lipids, and hormones which may be linked to proteins. Cow, human, horse, and fetal calves serum are usually employed (Mojica-Henshaw et al., 2013).

2.3 STEM CELL CULTURE

2.3.1 *STEM AND MATURE CELLS*

A stem cell is able to sustain its own population by self-renewal. It can propagate into one or several types of mature fully differentiated cells. Embryonic stem cells (ESCs) have a high proliferation rate since they must rapidly construct entirely a new organism. Adult stem cells are generally quiescent or proliferate slowly, but they can be mobilized upon demand, and induced for active proliferation, for example, hematopoietic stem cells (Kondo et al., 2003). The properties of stem cells are associated with their ability to perform symmetrical or asymmetrical division. Classical symmetrical divisions are operational in the self-renewal or expansion of the original stem cell pool. Asymmetrical division depends upon the existence of intracellular segregated microdomains and organization of genetic elements, as observed in oocytes. This organization allows micro-compartmentalized elements to be partially or totally segregated during cytoplasmic division in mitosis. Thus, a stem cell (mother cell) in the mitotic division can generate an identical unit and another one with a different profile (daughter cell), which is generally engaged and committed into differentiation along a required pathway.

2.3.2   STEM CELLS CAN BE CHARACTERIZED INTO THREE MAIN CLASSES

Embryonic, germinal, and somatic or adult stem cells are the main class of stem cells.

2.3.2.1   EMBRYONIC STEM CELLS (ESCS)

ESCs are derived from the early stage of an embryo, from morula, or the inner cell mass of blastocyst. Both are able to produce an entire organism. From morula, the source of cells can be derived from blastic cleavage, occurred just after the fertilization. Each cell (up to the 2 cell stage in humans, and 32 cell stage in animals) depending upon the species is able to generate a complete embryo. The cells of inner cell mass take part in the generation of all the tissues with the exception of extra-embryonic structure such as trophoblast and placenta. They are the main target of the worldwide research on ESCs.

2.3.2.2   THE GERMINAL STEM CELLS OR EMBRYONIC GERM CELLS (EGCS)

The germinal stem cells or EGCs (embryonic germ cells) are found in reproductive tissues during embryo development. They populate gonads that harbor progenitors of gametes. Similar cells are found in male adult gonads. Apparently, they have potential to generate multiple cell lineages, but the disadvantage of therapy is the difficulty in obtaining these cells, which would involve invasive surgery during embryo development.

2.3.2.3   SOMATIC STEM CELLS

Somatic stem cells are present in almost all the tissues (Jiang et al., 2002). The most well-known system is hematopoietic tissue. The hematopoietic system in adult is lodged within the bone marrow and its stem cells are specifically located in the endosteal region, close to internal bone surface. Moreover, bone marrow has at least three stem cell lineages: hematopoietic, mesenchymal, and endothelial stem cells. Among these, mesenchymal stem cells display broadest capacity of differentiation. Mesenchymal cells are associated with the inner cell mass and are not in direct contact with the external environment. Although these cells are most frequently of mesodermal origin, they can also be derived from ectoderm and endoderm, in a process termed as "epithelial-mesenchymal transition." Mesenchymal stem cells are able to restore tissue structural elements, called cellular stroma and the corresponding extracellular matrix. Mesenchymal stem cells are apparently pluripotent therefore; they can generate many cell types, with the possible exception of germline cells (Jiang et al., 2002).

2.3.2.4 CRYOPRESERVATION

Some of the cell types, for example, hESC, may need ultra-rapid freezing or vitrification (Hunt, 2011) where water is frozen in situ to form a glass and not permitted to infiltrate out of the cell as in slow freezing and is frequently utilized to freeze stem cells.

2.3.3 RESEARCH USING HUMAN EMBRYONIC STEM CELLS (HESCS)

The Human Fertilization and Embryology Act 1990 created the HFEA as an independent regulator of *in vitro* fertilization (IVF) and human embryo research. One of the statutory function of HFEA is to license and monitor establishments undertaking human embryo research. It includes the production of human embryonic stem cell (hESC) lines. The original act defined some purpose for which a research license could be issued by the HFEA:

• Endorsing advancements in the treatment of infertility;
• Increasing awareness about the causes of congenital disease;
• Increasing awareness about the causes of miscarriages;
• Developing more effective techniques for contraception.

2.3.4   STEM CELL LINES

Generally, the infectious threats that might arise with stem cell lines are not different from any others in which workers need to consider the probability of contamination with pathogens related with the species and tissue of origin. In the case of hESCs, the threat of contamination of the original donor tissue with the blood-borne pathogens is very low (Zou et al., 2004). However, once stem cell lines are differentiated to form tissue cell types they might offer a appropriate culture substrate for the growth of pathogenic viruses such as hepatitis C virus (HCV), hepatitis B virus (HBV) (Si-Tayeb et al., 2012), and other pathogens relying on the cell type created (Bandi and Akkina, 2008). Thus, when planning experiments to produce a particular differentiated cell type, thoughts should be given to the most probable contaminants that might arise in reagents, cells, and any test samples that might replicate in the differentiated cell type. Human iPSCs can be isolated from an extensive variety of tissues; therefore, the risk is associated with the tissue.

Human and mouse feeder cell lines applied to grow stem cells may also transmit viruses and can present parallel risks to those for continuous and finite cell lines. In addition, where primary mouse embryo fibroblasts (MEFs) are used to culture stem cells a range of viruses may occur in the original colony. Thus a viral screen for MEFs and mycoplasma testing should be performed.

2.3.5   STABILITY OF STEM CELL LINES

There are particular problems associated with cell lines resulting from stem cells, whether embryonic, fetal, and adult or induced pluripotent stem cells (iPSCs):

- Phenotypic stability is based on the medium, particularly on cytokines and the activity of specific pathways. For example, most mouse embryonic stem cell culture will tend to differentiate spontaneously unless the stem cell phenotype is maintained with LIF or a feeder layer.
- If allowed to differentiate, the resultant phenotype is regulated by regulatory factors in the medium, such as retinoic acid (RA)

or tetradecanoyl phorbol acetate (TPA), and the microenvironment (e.g., by the cell density, extracellular matrix, and signaling between cell types).

- It is not completely clear whether mesenchymal stem cells are integrally prone to genetic instability and transformation or are made so by genetic manipulation. Some reports claim that transformation does not occur, whereas others observed that it does (Ren et al., 2012), although this can be because of cross-contamination (Ren et al., 2011). Mesenchymal stem cells (often now termed mesenchymal stromal cells) generally do not have a clonal origin and this heterogeneity may explain differences in results. Pluripotent stem cell lines (i.e., human ESCs and human iPSCs), usually clonal in origin. They are well known to be susceptible to develop chromosomal changes and need to be periodically checked for genetic integrity. Again, it is rare cell in population that acquire a growth advantage, often at expense of the ability to differentiate, that take over the culture. Culture conditions that are suboptimal are especially prone to this. Re-cloning of cells and screening for sublines with normal karyotypes can work, otherwise reverting to an earlier passage is recommended.
- As for any other cell line, authentication of stem cell lines is also essential. The genotypic and phenotypic instability must be assessed.

2.3.6 KEY CONTRIBUTING FACTORS IN HUMAN PLURIPOTENT STEM CELLS (HPSC) CULTURE

Proliferation of mammalian cell *in vitro* needs growth media, extracellular matrix (ECMs), and environmental factors. In upcoming section, we will discuss three key features that affect the quality, robustness, and application of numerous hPSC culture methods: growth medium, extracellular matrices, and environmental cues (e.g., a growth environment in a bioreactor).

2.3.7 DEVELOPMENT OF GROWTH MEDIUM FOR HPSC CULTURE

Growth medium is the most critical parts of hPSC culture and has evolved dynamically since initially applied for hESC culture (Thomson et al., 1998). The first generation hESC medium usually comprised fetal bovine serum (FBS) and undefined/conditioned secretory constituents from mouse embryonic fibroblasts (MEFs). In current years, scientists have developed a more standardized and better-defined medium to substitute harmful elements in the media (Genbacev et al., 2005; Vallier et al., 2005). They are using chemically defined medium containing Activin A, Nodal, and FGF-2 to propagate hESCs. The knock-out serum replacement (KSR) is extensively applied with FGF-2 to help feeder-based hPSC culture. A defined culture medium (termed TeSR1) comprising FGF-2, lithium chloride (LiCl), g-aminobutyric acid (GABA), TGF-β, and pipecolic acid was created by Thomson and colleagues to apply in feeder-free circumstances (Ludwig et al., 2006). Lately, Thomson and coworkers created chemically definite E8 medium (E8), a derived from TeSR1 comprising eight components that are devoid of both serum albumin and b-mercaptoethanol. The E8 medium, in association with EDTA passaging, can be suitable for culturing a wide range of hiPSC and hESC lines, predominantly to increase episomal vector-based reprogramming efficiencies as well as experimental constancy (Chen et al., 2010, 2011).

2.3.8   SUSPENSION CULTURES PERMIT LARGE-SCALE PRODUCTION OF HPSCS

Many approaches established on colony culture, non-colony type of cell growth, and aggregated suspension cultures are applied to culture hPSCs. These approaches maintain the epithelial nature of hPSCs under the definite feeder, feeder-free, and xeno-free circumstances. Production of clinically relevant quantity of hPSCs, ranging from 10^7–10^{10} or more, is important for clinical use. Suspension culture in bioreactors offers a favorable platform for the robust manufacture of hPSC products. Usually, hPSCs expanded by the suspension culture in bioreactors remained pluripotent and stable at the chromosomal level. Among numerous kinds of bioreactors, spinner vessels and stirred-tank bioreactors are of particular interest.

KEYWORDS

- balanced salt solution
- embryonic stem cells
- fetal bovine serum
- g-aminobutyric acid
- human embryonic stem cell
- mouse embryo fibroblasts

REFERENCES

Aden, P., Paulsen, R. E., Mæhlen, J., Løberg, E. M., Goverud, I. L., Liestøl, K., & Lømo, J., (2011). Glucocorticoids dexamethasone and hydrocortisone inhibit proliferation and accelerate maturation of chicken cerebellar granule neurons. *Brain Res., 1418*, 32–41.

Bandi, S., & Akkina, R., (2008). Human embryonic stem cell (hES) derived dendritic cells are functionally normal and are susceptible to HIV-1 infection. *AIDS Res. Ther., 5*(1), 1–9.

Bryzek, A., Czekaj, P., Plewka, D., Tomsia, M., Komarska, H., Lesiak, M., Sieroń, A. L., Sikora, J., & Kopaczka, K., (2013). Expression and co-expression of surface markers of pluripotency on human amniotic cells cultured in different growth media. *Ginekologia Polska, 84*(12), 1012–1024.

Chaudhry, M. A., Bowen, B. D., & Piret, J. M., (2009). Culture pH and osmolality influence proliferation and embryoid body yields of murine embryonic stem cells. *Biochem. Eng. J., 45*(2), 126–135.

Chen, G., Gulbranson, D. R., Hou, Z., Bolin, J. M., Ruotti, V., Probasco, M. D., et al., (2011). Chemically defined conditions for human iPSC derivation and culture. *Nat. Methods, 8*(5), 424.

Chen, G., Hou, Z., Gulbranson, D. R., & Thomson, J. A., (2010). Actin-myosin contractility is responsible for the reduced viability of dissociated human embryonic stem cells. *Cell Stem Cell, 7*(2), 240–248.

Frahm, B., Blank, H. C., Cornand, P., Oelßner, W., Guth, U., Lane, P., Munack, A., Johannsen, K., & Pörtner, R., (2002). Determination of dissolved CO_2 concentration and CO_2 production rate of mammalian cell suspension culture based on off-gas measurement. *J. Biotechnol., 99*(2), 133–148.

Freshney, R. I., (2015). *Culture of Animal Cells: A Manual of Basic Technique and Specialized Applications.* John Wiley & Sons.

Genbacev, O., Krtolica, A., Zdravkovic, T., Brunette, E., Powell, S., Nath, A., Caceres, E., McMaster, M., McDonagh, S., Li, Y., & Mandalam, R., (2005). Serum-free derivation of human embryonic stem cell lines on human placental fibroblast feeders. *Fertil. Steril.*, *83*(5), 1517–1529.

Genzel, Y., König, S., & Reichl, U., (2004). Amino acid analysis in mammalian cell culture media containing serum and high glucose concentrations by anion exchange chromatography and integrated pulsed amperometric detection. *Anal. Biochem.*, *335*(1), 119–125.

Hunt, C. J., (2011). Cryopreservation of human stem cells for clinical application: A review. *Transfus. Med. Hemother.*, *38*(2), 107–123.

Jiang, Y., Jahagirdar, B. N., Reinhardt, R. L., Schwartz, R. E., Keene, C. D., Ortiz-Gonzalez, X. R., et al., (2002). Pluripotency of mesenchymal stem cells derived from adult marrow. *Nature*, *418*(6893), 41–49.

Jiang, Y., Vaessen, B., Lenvik, T., Blackstad, M., Reyes, M., & Verfaillie, C. M., (2002). Multipotent progenitor cells can be isolated from postnatal murine bone marrow, muscle, and brain. *Exp. Hematol.*, *30*(8), 896–904.

Kondo, M., Wagers, A. J., Manz, M. G., Prohaska, S. S., Scherer, D. C., Beilhack, G. F., Shizuru, J. A., & Weissman, I. L., (2003). Biology of hematopoietic stem cells and progenitors: Implications for clinical application. *Annu. Rev. Immunol.*, *21*(1), 759–806.

Kwong, P. J., Abdullah, R. B., & Khadijah, W. W., (2012). Increasing glucose in KSOM basal medium on culture day 2 improves *in vitro* development of cloned caprine blastocysts produced via intraspecies and interspecies somatic cell nuclear transfer. *Theriogenology*, *78*(4), 921–929.

Lee, W. Y., Park, H. J., Lee, R., Lee, K. H., Kim, Y. H., Ryu, B. Y., et al., (2013). Establishment and *in vitro* culture of porcine spermatogonial germ cells in low temperature culture conditions. *Stem. Cell Res.*, *11*(3), 1234–1249.

Ludwig, T. E., Levenstein, M. E., Jones, J. M., Berggren, W. T., Mitchen, E. R., Frane, J. L., Crandall, L. J., Daigh, C. A., Conard, K. R., Piekarczyk, M. S., & Llanas, R. A., (2006). Derivation of human embryonic stem cells in defined conditions. *Nature Biotechnol.*, *24*(2), 185–187.

Mojica-Henshaw, M. P., Morris, J., Kelley, L., Pierce, J., Boyer, M., & Reems, J. A., (2013). Serum- converted platelet lysate can substitute for fetal bovine serum in human mesenchymal stromal cell cultures. *Cytotherapy*, *15*(12), 1458–1468.

Onishi, T., Kinoshita, S., Shintani, S., Sobue, S., & Ooshima, T., (1999). Stimulation of proliferation and differentiation of dog dental pulp cells in serum-free culture medium by insulin-like growth factor. *Arch. Oral Biol.*, *44*(4), 361–371.

Ren, Z., Wang, J., Zhu, W., Guan, Y., Zou, C., Chen, Z., & Zhang, Y. A., (2011). Spontaneous transformation of adult mesenchymal stem cells from cynomolgus macaques *in vitro*. *Exp. Cell Res.*, *317*(20), 2950–2957.

Ren, Z., Zhang, Y. A., & Chen, Z., (2012). Spontaneous transformation of cynomolgus mesenchymal stem cells *in vitro*: Further confirmation by short tandem repeat analysis. *Exp. Cell Res.*, *318*(5), 435–440.

Ross, A. M., Nandivada, H., Ryan, A. L., & Lahann, J., (2012). Synthetic substrates for long-term stem cell culture. *Polymer*, *53*(13), 2533–2539.

Si-Tayeb, K., Duclos-Vallée, J. C., & Petit, M. A., (2012). Hepatocyte-like cells differentiated from human induced pluripotent stem cells (iHLCs) are permissive to hepatitis C virus (HCV) infection: HCV study gets personal. *J. Hepatol.*, *57*(3), 689–691.

Stryer, L., (1995). *Biochemistry* (4ᵗʰ edn.). W. H. Freeman and Company, New York.

Thomson, J. A., Itskovitz-Eldor, J., Shapiro, S. S., Waknitz, M. A., Swiergiel, J. J., Marshall, V. S., & Jones, J. M., (1998). Embryonic stem cell lines derived from human blastocysts. *Science*, *282*(5391), 1145–1147.

Vallier, L., Alexander, M., & Pedersen, R. A., (2005). Activin/nodal and FGF pathways cooperate to maintain pluripotency of human embryonic stem cells. *J. Cell Sci.*, *118*(19), 4495–4509.

Zou, S., Dodd, R. Y., Stramer, S. L., & Strong, D. M., (2004). Probability of viremia with HBV, HCV, HIV, and HTLV among tissue donors in the United States. *N. Engl. J. Med.*, *351*(8), 751–759.

CHAPTER 3

Bacterial Cell Culture

AMIT KUMAR SHARMA,[1] SANTOSH KUMAR MAURYA,[2] and
SHASHANK KUMAR[2]

[1]*Department of Agriculture, Uttaranchal University, Dehradun,
Uttarakhand, India, Tel.: +91 9793857582,
E-mail: amitbiochembhu@gmail.com*

[2]*School of Basic and Applied Sciences, Department of Biochemistry,
Central University of Punjab, Bathinda, Punjab–151001, India*

ABSTRACT

Bacterial culture (microbial culture) is a process of growing bacteria by allowing them to duplicate in prearranged media in controlled laboratory environments. Bacterial cultures are used to conclude the form of an organism or to check the abundance of the organism in the substance being tested. Occasionally it is required to isolate a pure culture of bacteria for study, diagnosis, or research purposes. Bacterial cultures are also used for storing the particular bacteria for further purposes.

3.1 INTRODUCTION

Some microorganisms are not visible under the microscope. It is not generally practical, to work with a single microorganism. For this reason, we study microbial cultures that contain a large number of microorganisms. A culture of a single bacterial species population is called axenic culture. Microbiologists customarily refer to such a culture as a pure culture. Although in a strictly technical sense a pure culture is one grown from, a single cell. If two or more than two species of microorganisms grow together, as they commonly do in nature, the mixed population is referred as a mixed culture.

As in animal cell culture, pure bacterial culture (cultures containing one species of the organism) is cultured regularly and sustained indefinitely by applying standard sterile techniques. Bacterial cells display an extensive degree of diversity in relation to nutritional and environmental requirements (Monod, 1942).

3.2   TYPE OF CULTURE

3.2.1   *STREAK CULTURE*

Streak culture of a microorganism is developed by drawing an inoculated needle in a straight line over the surface of the growth medium.

3.2.2   *STAB OR STICK CULTURE*

Stab culture is prepared in liquefied or solidified media using agar and gelatin. A pointed needle is inserted at some distance in the growth media to find out bacterial growth.

3.2.3   *STOCK CULTURE*

It is a culture of known microbial species maintained in the various labs for a long time. It is the stock of a single microorganism or its strains used for further study.

3.2.4   *STARTER CULTURE*

These cultures consist of some species or combination of species studied for the manufacturing of a particular product. For example, few laboratories are established to produce the milk product by *Lactobacillus*, *Leuconostoc*, and *Streptococcus* species. These bacteria can ferment milk products or perform some desirable changes according to our needs.

3.2.5   ENRICHMENT CULTURE

In this type of culture, the nutritional requirements of a bacteria (W) is increased in a media of mixed microbial population (W, X, Y, Z) to favor the growth of W bacteria. The nutritional environment is mediated by enrichment techniques. The flourished bacterial species W can be shifted to different media to obtain a pure culture.

3.2.6   BATCH CULTURE

In the batch culture, particular bacteria can be cultured in a small amount of liquid medium. Bacteria are inoculated from a source into a known culture volume and it starts to grow. This is applied to study the features of the growth of bacteria (Madigan et al., 2017).

3.3   SAFETY CONCERN FOR BACTERIAL CELL CULTURE

3.3.1   STERILIZATION

This process kills microorganisms and spores present in the media. All the apparatus used for culture should be sterilized in an autoclave. An autoclave is a simple pressure cooker like an instrument. Definite temperature (121°C) and pressure (1 bar) at a particular time period (15 minutes) kill the microorganisms present in the media.

3.3.2   ASEPTIC TECHNIQUES

The aseptic techniques used to reduce microbial contamination involves the disinfection of the working area, minimization of microbial contamination from the air in which the medium is to be open, and killing of microorganism by flame (Washington, 1996).

3.3.3   INOCULATION

It is a different method of introducing the bacteria into a growth media. Usually, the bacteria are introduced with the help of heat-sterilized loop and spread on the agar surface. The same technique is used for broth culture. We can also introduce bacteria into a Petri dish using pipettes.

3.3.4   INCUBATION

After bacterial inoculation, we transfer media to a special instrument known as incubator. We can adjust incubator's temperature of as required by the growing bacterial species. The culture of bacterial cells require much more precaution and sterile methods, we should also avoid contamination of pure cultures with different organisms. More importantly, precaution needs to be taken towards operator protection, especially from harmful microorganisms. Additionally, metallic apparatus utilized for the culturing techniques needs to be sterilized by heating. Furthermore, to prevent the bacterial spread, the workplace must be cleaned using the germicidal spray and/or UV radiation. Also with this precaution, all the instruments utilized in microbial cell culture need to be disposed of appropriately (Flickinger, 2010).

3.4   NUTRITION REQUIREMENT IN BACTERIA

Bacterial culture is different with animal cell culture due to bacterial diversity, the medium contents might be dissimilar and largely on the nutritional classification of organism. Bacteria generally fall into two classes, autotrophs (organisms that synthesize food in the form of sugar in the presence of sunlight as a source of energy) and heterotrophs (organisms that derive chemical energy through breaking down organic molecules). Autotrophs are further sub-grouped into chemoautotrophs or photoautotrophs or heterotrophs. Both chemo- and photoautotrophs depend on CO_2 as a carbon source but get energy from different sources. Chemoautotrophs utilize organic substances while the photoautotrophs are light-dependent. Chemo-heterotrophs and photo-heterotrophs both use the organic compound as a carbon source. Photo-heterotrophs use light (photon) for energy and the chemo subgroup acquires their energy from the metabolism of organic substances (Monod, 1949).

3.5   BACTERIAL CELL CULTURE MEDIA

3.5.1   *CULTURE MEDIA*

Nutrients present in solid or liquid form used for the growth, survival, and division of bacteria are called 'culture medium.' Bacterial culture media contain various nutrients that promote the growth of respective bacteria. Some media have both inorganic and organic contents. Plant extracts and animal tissue are utilized for preparation bacterial culture media.

Numerous kind of medium are applied in bacterial culture categorized as complex or defined media. The former generally comprises natural substances, containing meat and yeast extract. They are not very well defined, as their exact constituents are largely unknown. These media are rich in nutrients usually appropriate for culture of fastidious bacteria that requires a mixture of growth nutrients.

3.5.2   *DEFINED MEDIA*

Defined media are relatively simple. They are designed according to specific need of bacterial species that is cultivated. They are consisting of known constituents in the essential amounts. This suppleness is generally utilized to choose or remove particular species by taking benefit of their distinguished nutritional requirement. For example, bile salts can be incorporated in the media to select enteric bacteria (rod-shaped Gram-negative bacteria like *Salmonella* or *Shigella*) since the growth of other Gram-negative and Gram-positive bacteria will be repressed (Kenneth et al., 2014).

3.6   CLASSIFICATION OF MEDIA

3.6.1   *CLASSIFICATION OF MEDIA ON THE BASIS OF COMPOSITION*

These types of media do not require living cells or tissue. They may be again grouped into two subtypes, i.e., nonsynthetic media and synthetic media.

3.6.1.1   NON-SYNTHETIC MEDIA

Constituents and composition of this media are not known. Potato-Dextrose-Agar (GM-25), Oatmeal- Agar (GM-24), Soil-Extract-Agar (SM-1), Waksman's medium (GM-40), Malt-Extract-Agar (GM- 19b) are some of the generally used non-synthetic media.

3.6.1.2   SYNTHETIC MEDIA

Constituents and composition of this type of media are exactly known and are used for metabolic and nutritional studies. For example, Czakek's-Dox Medium (GM-9) and Richard's solution (GM-27), etc.

3.6.2   CLASSIFICATION OF MEDIA ON THE BASIS OF PHYSICAL STATE

3.6.2.1   LIQUID MEDIA

They are present in liquid form and used for culture purposes for examples-Nutrient, Skimmed, Broth, Peptone Solution, Milk, etc.

3.6.2.2   SEMISOLID MEDIA

A little amount (less than 0.5%) of agar is present in this media imparts a "custard consistency," for example, include Cystine Trypticase Agar Medium, etc.

3.6.2.3   LIQUEFIABLE SOLID MEDIA

This medium is also called as solidly reversible to the liquid medium. Use of agar and gelatin makes this media solid in warm and liquid in cold conditions, for example, Nutrient-Agar Medium, Nutrient- Gelatin Medium, and Potato-Dextrose-Agar Medium, etc.

3.6.2.4   SOLID MEDIA

This type of media always remains in solid-state for example coagu-lated blood serum, coagulated egg, potato slices, trypticase-soy-agar medium, etc.

3.6.3   CLASSIFICATION OF MEDIA ON THE BASIS OF FUNCTION

3.6.3.1   CULTIVATION MEDIA

This type of media is used for bacterial cultivation, for example, Nutrient Agar, nutrient broth, etc.

3.6.3.2   STORAGE MEDIA

Stock culture is applied to store bacteria for a long time to provide a stock of viable cultures. The medium used for the stock culture is called a storage medium, for example, yeast extract mannitol agar medium, etc.

3.6.3.3   ENRICHMENT MEDIA

This type of media is used to promote the growth of selected bacteria by adjusting the nutritional environment. The media is helpful for scientists to identify and isolate selected bacteria among their mixed populations. For example, the addition of animal tissue and plant extracts to the agar culture media provides essential elements for the culturing of bacteria in culture disk.

3.6.3.4   DIFFERENTIAL MEDIA

This media can identify the presumptive bacterial species in culture. The best example is Blood Agar Medium. If we mix a population of various bacteria into a blood agar medium, some will hemolyze and form a zone and some will not.

3.6.3.5   ASSAY MEDIA

Certain media can influence the bacterial cells to form antibiotics, toxins, and enzymes other products and are regarded as assay media or media for specific purposes. For example, the pyridoxine-deficient culture medium for *Streptococcus faecalis* produces cells containing a large amount of tyrosine decarboxylase apo-enzyme.

3.6.3.6   MAINTENANCE MEDIA

To maintain physiological and viable characters of the cultured bacteria there is a requirement of optimized maintenance medium for the optimum growth of bacteria. The fast and optimized growth of bacteria leads to a fast death rate simultaneously. For example, glucose in a media promotes growth as well as death. So, exclusion of glucose is preferable in the maintenance medium (Wilson and Miles, 1965).

3.7   CULTURE PROCEDURES FOR BACTERIAL CELLS

In small-scale culture, bacteria are grown in laboratories by utilizing either solid or liquid media. Liquid media are generally prepared in flasks and inoculated with an aliquot of organism to be cultured. The media is then agitated constantly on a shaker. The procedure mixes and ensures the cultures in suspension form. For these cultures, enough space needs to be permitted above the medium to enable suitable diffusion of O_2 into the solution. Therefore, as a rule of thumb, the volume of culture medium added to the flasks should not surpass more than 20% of the total volume of the flask. This is very important for aerobic bacteria and less anaerobic microorganisms (Hinshelwood, 1946).

In industrial cultures, bioreactors or fermenters fitted with the stirring device to improve mixing also gas interchange might be utilized. The device usually contains probes that control changes in oxygen, pH, concentration, and temperature. Also, most of the systems are encircled by a water jacket with flowing cold water to lower the heat produced during fermentation. Outlets are also built to release CO_2 and other gases generated by cell metabolism. During use of fermenters, it is precaution should take to lower the potential contamination of airborne microorganisms when air is bubbled via the culture. Sterilization of air therefore, is necessary and can be attained by using a filter (pore size of approximately 0.2 mm) at the air entry point in the chamber. Solid medium are generally prepared by solidifying the selected medium with 1–2% of the seaweed extract agar. The composition is microbe-resistant and thereby provides an inert gel medium for bacterial growth. Solid agar media are commonly utilized to separate mixed culture and isolation of pure bacterial cultures. This is done by streaking diluted culture of bacteria at the surface of agar plate utilizing a sterile inoculating loop. Cells streaked on the plate will ultimately proliferate into a colony. Every colony is produced by same species of bacteria. Once isolated, bacteria might be cultivated either in batch or continuous cultures. Batch cultures are commonly utilized for liquid growth and entail aliquot inoculation of cells into a sterile flask comprising finite amount of medium. These systems are called closed because nutrient supply is restricted at the starting point of culture. In these conditions, growth will continue until the nutrients in medium are depleted or build-up of toxic waste materials. In such systems, the cellular composition and physiological status of the cell will diverge throughout the whole growth cycle. In continuous culture (also referred to as open system), the medium is changed frequently to change that spent by the cells. The aim of this system is to keep the cells in the exponential phase. Nutrients, biomass, and waste products are regulated by changing the dilution rate of culture. Continuous culture is more complex to set up but has advantage over batch cultures. They enable the growth under steady-state condition which guarantees a strong coupling between biosynthesis and cell division. As a consequence, the physiological features of the cultures are clearly defined with small changes in the cellular composition of the cells during growth cycle. The main problem with the open system is increased risk of contamination related with the dilution of culture. However, by applying of strict aseptic techniques while feeding or harvesting of cells may lower the possibility of contaminations. In addition, the whole system can be automated by connecting the culture vessels to reservoirs via solenoid valves that can be triggered to open when needed. This minimizes the direct contact with outside environment and contamination also (Droop, 1966).

3.8 BACTERIAL GROWTH CULTURE DETERMINATION

Hemocytometer is needed to conclude the growth of bacteria in culture as it count cells directly. Cells are cultured on solid agar plates and counting of colony is used to find out the growth. This approach assumes that every colony is formed from a single cell that might not always be the case, because errors in dilution and/or streaking results in clumps instead of single-cell colonies. Furthermore, suboptimal culture conditions might cause poor growth resulting to the underestimation of cell count. When cells are cultured in suspension, alteration in turbidity of the growth medium could be determined by spectrophotometer. The absorbance value from the spectrophotometer is converted to cell number utilizing a standard curve of absorbance versus cell number. This should be constructed for each and every cell type by getting readings of a series of known number of cells in the media (Amann et al., 1995).

3.9   POTENTIAL USE OF BACTERIAL CELL CULTURES

- Various microbial and animal cell cultures are gradually used not only by scientists (to study the activity of cells in isolation) but also by many biotechnology and pharmaceutical companies to produce biological product such as antibodies (e.g., OKT3 used in suppressing immunological organ rejection in transplant surgery) viral vaccines (e.g., polio vaccine), and recombinant proteins.
- Microbial culture is applied to amplify recombinant DNA in bacteria using bacterial culture and led to an ever-expanding list of better-quality products, both from mammalian and bacterial cells, for the therapeutic use. These products comprise commercial production of blood clotting factor VIII for hemophilia, insulin for diabetes, interferon a and b for anticancer chemotherapy and erythropoietin (EPO) for anemia.
- Bacterial culture are commonly applied for other industrial uses such as the large- scale production of cell growth regulators, proteins, alcohols, organic acids, solvents, sterols, surfactants, amino acids, vitamins, etc. (Koch, 1998).
- Microbial cells are applied for the degradation of waste products mainly from agricultural and food industries.
- Bacteria are applied for to bioconversion of waste material into beneficial end products. In a preliminary toxicological study of xenobiotics, these organisms are quickly substituting animals (Davis et al., 2005).

KEYWORDS

- bacterial cell culture
- incubation
- liquefiable solid media
- starter culture
- sterilization
- streak culture

REFERENCES

Amann, R. I., Ludwig, W., & Schleifer, K. H., (1995). Phylogenetic identification and *in situ* detection of individual microbial cells without cultivation. *Microbiol. Rev.*, *59*(1) 143–169.

Davis, K. E., Joseph, S. J., & Janssen, P. H., (2005). Effects of growth medium, inoculum size, and incubation time on culturability and isolation of soil bacteria. *Appl. Environ. Microbiol.*, *71*(2), 826–834.

Droop, M. R., (1966). Organic acids and bases and the lag phase in *Nannochloris oculata*. *J. Mar.Biol. Assoc. U.K.*, *46*(3), 673–678.

Flickinger, M. C., (2010). *Encyclopedia of Industrial Biotechnology: Bioprocess, Bioseparation, and Cell Technology* (Vol. 7).

Hinshelwood, C. N., (1946). *Chemical Kinetics of the Bacterial Cell*. Oxford at the Clarendon Press, London.

Kenneth, J. R., Ryan, C., & Ray, G., (2014). *Sherris Medical Microbiology*. McGraw-Hill Education/Medical.

Koch, A. L., (1998). Microbial physiology and ecology of slow growth. *Microbiol. Mol. Biol. Rev.*, *62*(1), 248.

Madigan, M. T., Martinko, J. M., & Parker, J., (2017). *Brock Biology of Microorganisms* (Vol. 13).Pearson, Upper Saddle River, NJ: Prentice Hall.

Monod, J., (1942). *Research on the Growth of Bacterial Cultures*. Hermann & Cie, Paris. 211.

Monod, J., (1949). The growth of bacterial cultures. *Annu. Rev. Microbiol.*, *3*(1), 371–394.

Washington, J. A., (1996). Principles of diagnosis. *Medical Microbiology*. University of Texas Medical Branch at Galveston.

Wilson, G. S., & Miles, A. A., (1965). Topley and Wilson's principles of bacteriology and immunity. *Academic Medicine*, *40*(3), 317.

CHAPTER 4

Potential Use of Cell Cultures

AMIT KUMAR SHARMA,[1] ATUL KUMAR SINGH,[2] HADIYA HUSAIN,[3] and SHASHANK KUMAR[2]

[1]Department of Agriculture, Uttaranchal University, Dehradun, Uttarakhand, India, Tel.: +91 9793857582, E-mail: amitbiochembhu@gmail.com

[2]School of Basic and Applied Sciences, Department of Biochemistry, Central University of Punjab, Bathinda, Punjab–151001, India

[3]Biochemical and Genetics Research Lab, Section of Genetics, Department of Zoology, Faculty of Life Science, Aligarh Muslim University, Aligarh–202002, Uttar Pradesh, India

ABSTRACT

Substances with biological importance need to be tested in living systems due to its effect on living organisms. Nowadays, it's a trend to avoid the chemical substances testing directly to the animals due to ethical issues. With the cell cultures approach, it is possible to determine the biological activity and the mechanism of action of biologically important compounds straight at the cellular level. Cultures of cells derived from various tissues and organs and have the potential to show the effects of chemicals. These tests can thus distinguish negative erratic effects on individual tissues and probable sensitivity in a microorganism.

4.1 INTRODUCTION

Cell culture is a key technique applied in molecular and cellular biology. It enables the exceptional model system for the learning of usual cellular biochemistry and physiology (e.g., aging, metabolic studies), the effect of toxic compound, carcinogenesis, and mutagenesis. It is applied in the screening and development of drugs and broad-scale production of biologically important compounds (e.g., therapeutic proteins, vaccines). The main benefit of cell culture for these applications includes constancy and reproducibility of outcome. It could be obtained from a batch of clonal cells.

4.1.1 APPLICATIONS IN SOLVING BIOLOGICAL PROCESS

Considerable knowledge about cellular structure and function is possible due to cell culture technique. This includes intracellular synthetic process of macromolecules, like proteins and nucleic acids and other relevant cellular molecules. Energy and signal transmission inside the cell, cytoskeletal architecture, and transport through membrane, product formation, cell-cell interaction, malignant transformation, and microbial infections are characteristically studied in cell cultures. A few examples have been established for its utilization in research and production. This focuses on the significance of cell aggregates grown in a three-dimensional (3D) space, a method that might become a standard technique in future.

4.1.2 APPLICATION OF CELL CULTURE TO DETECT VIRUSES IN THE ERA OF TECHNOLOGY

Viral disease identification is conventionally depended on the isolation of the pathogenic virus in cell culture. Although this tactic is generally, slow and needs substantial technical efficiency. It has been considered for a long time as a "gold standard" for laboratories to diagnose viral disease. With the introduction of nonculture methods for the rapid detection of viral antigens and/or nucleic acids, the practicality of viral culture has been criticized. The assessment pronounced improvements in cell culture-based viral diagnostic products and procedures, with the utilization of fresh cell culture arrangements, cryopreserved cell culture, centrifugation-improved inoculation, co-cultivated cell culture, precytopathogenic effect detection, and transgenic cell lines. These methods provide more effective and technically less challenging viral identification in cell culture. Though maximum laboratories follow numerous cultures and nonculture tactics to improve viral disease diagnosis but isolation of viruses in cell culture is still a useful tactic, particularly when the isolate desired is viable. Besides these techniques are also used to distinguish viable and nonviable viruses in a typical viral disease. Culture-based techniques can deliver a more appropriate result than molecular methods.

4.1.3   APPLICATION IN FUNDAMENTAL CELL CULTURE RESEARCH

Intracellular activity in elementary research includes DNA transcription, translation with labeled radioactive isotopes or fluorescence, precise cell lines, assays for cell cycle and senescence, characterization, proliferation, differentiation, apoptosis, and metabolic analysis by immobilization techniques (Hoffman et al., 1998; Andreeff et al., 2009; Kourtis and Tavernarakis, 2009). In addition, analysis of biomolecules intracellular flow such as RNA processing, protein transport, microtubule assembly, and disassembly, permits establishment of dedicated cell lines for the research purpose. Info genomics and proteomics, this method is utilized for genetic analysis, metabolic pathways, gene expression, infection, immortalization, cellular transformation, cellular cooperation, proliferation, morphogenesis, adhesion, and ECM interactions between host and parasite (Andreeff et al., 2009).

4.1.4   APPLICATION OF CELL CULTURE IN APPLIED RESEARCH

The cell culture techniques are applied to study virology and for the isolation of viruses. In this technique animal are used as they provide large number of cells (appropriate for virus isolation), facilitated prevention of contamination by using antibiotics, clean-air apparatus, and decrease the number of required experimental animals (Lindenbach et al., 2005; Leland and Ginocchio, 2007). Moreover, plant cell culture (Mokili et al., 2012), vaccine production, and biotechnological analysis of drug production in bioreactors (insulin, growth hormones, interferon, etc.) have to be made (Rates, 2001; Li et al., 2010). Otherwise, applications of cell culture in pharmacology and toxicology techniques include determination of the effect of various drugs, interaction type in drug-receptor, resistance phenomena, cytotoxicity, mutagenesis, and carcinogenesis. One field of application, reasonably studied in current years is the tissue engineering, e.g., *in vitro* production of tissues as cartilage or skin for the treatment of burns and induced differentiation autografting (Naderi et al., 2011).

4.1.5 APPLICATION OF CELL CULTURE IN CANCER RESEARCH

Applications of cell culture in the development of anticancer drugs have been very valuable and novel techniques. For example utilization of 3D models of cell culture (Smalley et al., 2008) either anchorage-independent (i.e., non-adherent) or anchorage-dependent (adherent to a substrate). The accumulation of cells could be attained by utilizing low-attachment plates and by coating of surface in 3D cultures without anchorage (for example, poly-hydroxyethyl methacrylate, and agarose) (Friedrich et al., 2007; Lovitt et al., 2014).

3D environment in the model could be created by utilizing pre-fabricated scaffolds. These scaffolds anchorage-dependent and contain porous materials to assist the growth of 3D structures. They are referred as multilayered cell culture (MCCs) and are outcome of cells attached to specific substrates made up of tumor cells. They are grown on a membrane particularly designed to permit the measurement of drug diffusion.

Establishment of microfluidic channels could also be accomplished to support the 3D cell culture additionally; ECM is also supplied inside these chambers to permit ECM-to-cell collaboration (Lovitt et al., 2014; Toh et al., 2007). The advantage of cells grown-up in 3D culture environment includes oxygen and nutrient gradients mediated amplification of cell-to-cell interaction, diverse degree of cellular proliferation, etc. (Lovitt et al., 2014). The main aim of generating 3D cell culture arrangements differ extensively from the designing of tissues for clinical delivery and creation of models for drug screening (Haycock et al., 2011).

4.1.6 APPLICATION IN TISSUE ENGINEERING

Definitely one of the greatest developments in cell culture technique is the development of tissue engineering. It enables *in vivo* designing and the establishment of cell lines. In the future, this can be used as a tissue alternative to treat damaged or malfunctioned tissue in the patients. Regardless of partial accomplishment in a few complicated organs, the potential of substituted tissue has been achieved for few targets. The clinical accomplishments in skin (Lazic and Falanga, 2011), cartilage (Rodríguez-Hernández et al., 2014), bladder (Atala, 2009), and trachea (Macchiarini et al., 2008) revealed tissue engineering as a major achievement in biomedical field (Zorlutuna et al., 2013).

Synthesis of artificial ECM assembly centered on ECM components or synthetic ingredients is another area of tissue engineering. It offers techniques for the development of cellular microenvironment (Zorlutuna et al., 2013). Thus, tissue engineering is nowadays deemed as end product for regenerative medicine and also permits technology for different areas of research from drug discovery to the promising biorobotics area (Zorlutuna et al., 2013). This revolutionary technique encompasses the development of new notions and cell culture technology. In traditional methods of cell culture, tissue development relies on a 3D cell establishment and development of a suitable extracellular matrix (ECM) (Sittinger et al., 1996). It highlights the development of matrix and cell differentiation in a synthetic tissue. Furthermore, distinct constituents of ECM have been extensively utilized as a platform material for tissue engineering (Zorlutuna et al., 2013). Applications like 3D tissue models to examine drugs (to assess anticancer drug efficacy, toxicity, monitoring events such as proliferation and apoptosis, cellular immunity by cytokine release, among others, therapeutic drug assessment) in the advancement of multicellartificial cornea imitate (Vrana et al., 2008; Götz et al., 2012), cancer models, and in biorobotics by utilizing cell-based arrangements as "actuators" (e.g., cardiomyocytes, vorticella, myoblast) with the capability to react piezoelectric materials (Zorlutuna et al., 2013; Ueda et al., 2010) has been utilized for the potential of tissue engineering.

4.1.7 APPLICATIONS IN TOXICOLOGY AND PHARMACOLOGY

Establishment of cell lines permits toxicological and pharmacological studies. The main objectives of these studies to know the effect of metal concentrations on the morphological variations of cells. Main objective is to assess the effect of drug on metabolism and defense system of cell (Fuente et al., 2002; Carranza-Rosales et al., 2005). These parameters are concerned with therapeutic outcome of patients (Gutiérrez et al., 2007).

4.1.8 MISCELLANEOUS CELL CULTURE APPLICATION

In the era of advance molecular biology, cell biology, biochemistry, and other areas of biological studies, it is possible to present new methods for the establishment and characterization of new cell lines. These cells can be used to perform studies on cell interaction, host-parasite interaction, and improvement of organ function affected by disease or trauma, and so on. Animal cell culture technology has progressed meaningfully since previous few decades and is considered as a dependable, healthy, and comparatively matured technology (Li et al., 2010).

Meanwhile, the regularization of cell culture has created a technological revolt. It involves genetic manipulation of cell lines, rich media development, vaccine design, development of progressively more sophisticated incubators for preferred atmosphere, development of 3D scaffolds, new adhesion surface, robotization process, etc. Furthermore, a variety of biotherapeutics has been synthesized using cell culture approaches at a broad-scale. It facilitates the profitable use of applied technology and clinical as well as research studies (Li et al., 2010).

KEYWORDS

- cell culture
- extracellular matrix
- multilayered cell culture
- pharmacology
- tissue engineering
- toxicology

REFERENCES

Andreeff, M., Goodrich, D. W., & Pardee, A. B., (2009). Cell proliferation, differentiation, and apoptosis. *Cancer Medicine* (5th edn).

Atala, A., (2009). Engineering organs. *Curr. Opin. Biotechnol., 20*(5), 575–592.

Carranza-Rosales, P., Said-Fernández, S., Sepúlveda-Saavedra, J., Cruz-Vega, D. E., & Gandolfi, A. J., (2005). Morphologic and functional alterations induced by low doses of mercuric chloride in the kidney OK cell line: Ultrastructural evidence for an apoptotic mechanism of damage. *Toxicology, 210*(2/3), 111–124.

Friedrich, J., Ebner, R., & Kunz-Schughart, L. A., (2007). Experimental anti-tumor therapy in 3-D: Spheroids-old hat or new challenge? *Int. J. Radiat. Biol., 83*(11/12), 849–874.

Fuente, H. D. L., Portales-Perez, D., Baranda, L., Diaz-Barriga, F., Saavedra-Alanis, V., Layseca, E., & Gonzalez-Amaro, R., (2002). Effect of arsenic, cadmium, and lead on the induction of apoptosis of normal human mononuclear cells. *Clin. Exp. Immunol., 129*(1), 69–77.

Götz, C., Pfeiffer, R., Tigges, J., Blatz, V., Jäckh, C., Freytag, E. M., Fabian, E., Landsiedel, R., Merk, H. F., Krutmann, J., & Edwards, R. J., (2012). Xenobiotic metabolism capacities of human skin in comparison with a 3D epidermis model and keratinocyte-based cell culture as *in vitro* alternatives for chemical testing: Activating enzymes (Phase I). *Exp. Dermatol., 21*(5), 358–363.

Gutiérrez, G., Mendoza, C., Zapata, E., Montiel, A., Reyes, E., Montaño, L. F., & López-Marure, R., (2007). Dehydroepiandrosterone inhibits the TNF-alpha-induced inflammatory response in human umbilical vein endothelial cells. *Atherosclerosis, 190*(1), 90–99.

Haycock, J. W., (2011). *3D Cell Culture: A Review of Current Approaches and Techniques* (pp. 1–15). In 3D, cell culture. Humana Press.

Hoffman, B. B., Sharma, K., Zhu, Y., & Ziyadeh, F. N., (1998). Transcriptional activation of transforming growth factor-β1 in mesangial cell culture by high glucose concentration. *Kidney Int., 54*(4), 1107–1116.

Kourtis, N., & Tavernarakis, N., (2009). Cell-specific monitoring of protein synthesis *in vivo*. *PLoS One., 4*(2), e4547.

Lazic, T., & Falanga, V., (2011). Bioengineered skin constructs and their use in wound healing. *Plast. Reconstr. Surg., 127*, 75S–90S.

Leland, D. S., & Ginocchio, C. C., (2007). Role of cell culture for virus detection in the age of technology. *Clin. Microbiol. Rev., 20*(1), 49–78.

Li, F., Vijayasankaran, N., Shen, A., Kiss, R., & Amanullah, A., (2010). Cell culture processes for monoclonal antibody production. *mAbs., 2*, 466–477.

Lindenbach, B. D., Evans, M. J., Syder, A. J., Wölk, B., Tellinghuisen, T. L., Liu, C. C., et al., (2005). Complete replication of hepatitis C virus in cell culture. *Science, 309*(5734), 623–626.

Lovitt, C. J., Shelper, T. B., & Avery, V. M., (2014). Advanced cell culture techniques for cancer drug discovery. *Biology, 3*(2), 345–367.

Macchiarini, P., Jungebluth, P., Go, T., Asnaghi, M. A., Rees, L. E., Cogan, T. A., Dodson, A., Martorell, J., Bellini, S., Parnigotto, P. P., & Dickinson, S. C., (2008). Clinical transplantation of a tissue-engineered airway. *Lancet, 372*(9655), 2023–2030.

Mokili, J. L., Rohwer, F., & Dutilh, B. E., (2012). Metagenomics and future perspectives in virus discovery. *Curr. Opin. Virol., 2*(1), 63–77.

Naderi, H., Matin, M. M., & Bahrami, A. R., (2011). Critical issues in tissue engineering: Biomaterials, cell sources, angiogenesis, and drug delivery systems. *J. Biomater. Appl., 26*(4), 383–417.

Rates, S. M. K., (2001). Plants as source of drugs. *Toxicon., 39*(5), 603–613.

Rodríguez-Hernández, C. O., Torres-García, S. E., Olvera-Sandoval, C., Ramírez-Castillo, F. Y., Muro, A. L., Avelar-Gonzalez, F. J., & Guerrero-Barrera, A. L., (2014). Cell culture: History, development and prospects. *Int. J. Curr. Res. Aca. Rev., 2*(12), 188–200.

Sittinger, M., Bujia, J., Rotter, N., Reitzel, D., Minuth, W. W., & Burmester, G. R., (1996). Tissue engineering and autologous transplant formation: Practical approaches with resorbable biomaterials and new cell culture techniques. *Biomaterials, 17*(3), 237–242.

Smalley, K. S., Lioni, M., Noma, K., Haass, N. K., & Herlyn, M., (2008). *In vitro* three-dimensional tumor microenvironment models for anticancer drug discovery. *Expert Opin. Drug Discov., 3*(1), 1–10.

Toh, Y. C., Zhang, C., Zhang, J., Khong, Y. M., Chang, S., Samper, V. D., Van, N. D., Hutmacher, D. W., & Yu, H., (2007). A novel 3D mammalian cell perfusion-culture system in micro fluidic channels. *Lab Chip, 7*(3), 302–309.

Ueda, J., Secord, T. W., & Asada, H. H., (2010). Large effective-strain piezoelectric actuators using nested cellular architecture with exponential strain amplification mechanisms. *IEEE ASME Trans. Mechatron., 15*(5), 770–782.

Vrana, N. E., Builles, N., Justin, V., Bednarz, J., Pellegrini, G., Ferrari, B., Damour, O., Hulmes, D. J., & Hasirci, V., (2008). Development of a reconstructed cornea from collagen-chondroitin sulfate foams and human cell cultures. *Invest. Ophthalmol. Vis. Sci., 49*(12), 5325–5334.

Zorlutuna, P., Vrana, N. E., & Khademhosseini, A., (2013). The expanding world of tissue engineering: The building blocks and new applications of tissue engineered constructs. *IEEE Rev. Biomed. Eng., 6*, 47–62.

CHAPTER 5

Principles of Clinical Biochemistry and Biochemical Analysis

SHASHANK KUMAR

School of Basic and Applied Sciences, Department of Biochemistry, Central University of Punjab, Bathinda, Punjab–151001, India, Tel.: +91 9335647413, E-mail: shashankbiochemau@gmail.com

ABSTRACT

Clinical biochemistry is a discipline which usually deals with the examination of body fluids such as blood, serum, and urine for therapeutic and diagnostic purposes. It is an applied form of biochemistry which involves the use and measurement of enzyme activities by spectrophotometry, electrophoresis, and immunoassay method. This branch initiated in the late 19th century with the use of basic chemical reaction tests for numerous constituents of blood and urine. Nowadays, there are many blood and urine tests are available with extensive diagnostic abilities.

5.1 INTRODUCTION

5.1.1 PRINCIPLES OF CLINICAL BIOCHEMISTRY

Chemistry is a division of science which deals with the understanding of matter. The chemistry that deals with biological matter are known as Biochemistry. Biochemistry dealing with the human health and diseases is known as medical biochemistry (Figure 5.1). Clinical biochemistry is also referred to as clinical chemistry/chemical pathology or medical biochemistry (Figure 5.1). Basically, it is a branch of clinical pathology generally related to the understanding of body fluids diagnosis as well as therapeutic purpose. Clinical biochemistry is an application aspect of medical biochemistry. It deals methodology and interpretation of clinical (chemical) tests carried out for the diagnosis of disease.

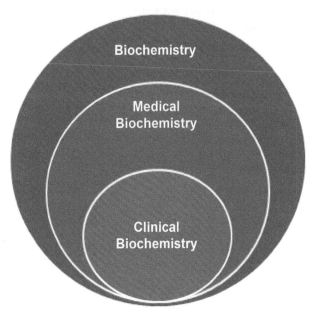

FIGURE 5.1   The relationship between biochemistry, clinical biochemistry, and medical biochemistry.

Protein (amino acids), carbohydrate (sugars) and lipids (fatty acids and glycerol) are the chemical component of the human body (Figure 5.2). Besides DNA and RNA, the nucleic acids residing in the body are formed of nucleotides/nucleosides. For life, energy is the main requirement and is generated by the oxidization of food material we eat. Energy is produced in the form of ATP in the body. The component (plasma membrane, chromosomes, etc.) of the cells are comprised of different macromolecule (protein, carbohydrates, lipids, and nucleic acids). The food we take, is utilize for energy production and synthesis of cellular component (extracellular and intracellular) with the help of enzymatic anabolic and catabolic reactions. Body systems such as digestion, circulation, excretion, and nervous system utilize these reactions to perform various functions such as maintenance of the components of blood and plasma, enzyme synthesis and degradation, transportation of particles across the cell membrane, synthesis, and action of hormones, blood coagulation, signal transduction, etc. In a normal individual, the metabolism process (anabolism and catabolism) goes smoothly and the metabolites circulating through the biological fluids are in normal concentration. During a disease condition, the homeostasis between anabolism and catabolism gets altered and persists for a small/long time in the body.

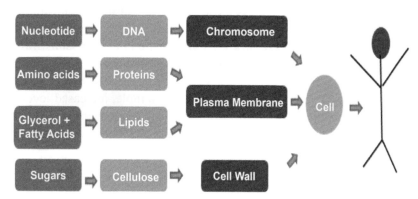

FIGURE 5.2   Integration of biomolecules.

The alteration generates the alterations in metabolite and concentration of few elements in biological fluids and inside the body. Thus the human biochemistry is associated with the learning of normal and healthy human body, however, normal biochemistry (metabolic/physiological/biochemical processes) is altered in several diseases and clinical conditions such as albinism, diabetes mellitus, night blindness, thalassemia, kidney dysfunction, atherosclerosis, rickets, liver disease, jaundice, SCA (sickle cell anemia), etc. Clinical biochemistry deals with the study of aberrant biochemistry of the human body in diseased conditions. It helps to understand the biochemical basis of many diseases at diagnosis, prognosis, and treatment level. Following tests/measurements helps in the identification of physiological, pathological, nutritional, and genetic diseases:

- **Assay for of Serum Diagnostic Enzymes:** For example, assay of serum alkaline phosphatase (ALP) helps in the detection of osteoblastic bone disease and obstructive jaundice.
- **Estimation of Normal/Abnormal Concentration of Metabolite:** For example, the presence of galactose and fructose in urine

indicates the occurrence of galactosemia and fructose intolerance respectively.

- **Assay for of Hormones:** For example, T_3, T_4, TSH, growth hormone, insulin, and several other hormones assays help in the diagnosis of endocrine disorder such as hypothyroidism, thyrotoxicosis, diabetes mellitus, dwarfism, Cushing's disease, etc.
- **Estimation of Electrolyte Concentration:** For example, quantitative estimation of Na^+, K^+, Ca^{++} and Mg^{++} in the blood sample is useful to identify skeletal muscles, heart, and bone disorders.
- **Quantitative and Qualitative Examination of Urine Constituent:** For example, Bence Jones protein present in urine indicates multiple myeloma.
- **Various Organ Function Tests:** For example, increased levels of serum creatinine, blood urea, and BUN. The presence of protein in urine and decreased renal clearance indicate kidney dysfunction.

Overall, the above discussion shows that clinical biochemistry is an important major branch of pathology.

5.2 PRINCIPLES INVOLVE IN CLINICAL BIOCHEMICAL ANALYSIS

5.2.1 BASIC CONCEPTS

Reliability of quantitative and qualitative clinical result analyses needs knowledge of the following domains:

- Solvent and solute concept;
- Units of measurement;
- Chemicals and reference material;
- Basic techniques;
- Safety.

5.2.1.1 WAYS TO EXPRESS SOLUTION CONCENTRATION

A solution is a combination of one/more solute dispersed in dissolving solvent (gaseous, liquid, or solid) homogeneously. In laboratory practice, solutes are typically referred to as analyte. In the clinical facilities, mostly the analysis involves the determination of substance concentration in solutions (blood, serum, urine, spinal fluid, or other body fluids). The international system of units (SI) endorses the utilization of moles to represent solute concentration in a unit volume of solution to represent the analyte concentration, and the utilization of liter to refer the volume. There are different means to represent the percentage of solute in a solvent. Percentage weight by weight (% w/w): Percent in this regard signifies the number of grams of solute present in 100 grams of solution. As an example, let's think through a 12% by weight sodium chloride solution. Such a solution would contain 12 grams of NaCl in every 100 grams of solution. To make such a solution, you could weigh out 12 grams of NaCl, and add with 88 grams of solvent water, thus the resulting mass of the solution is 100 grams. Volume by volume percentage (% V/V): in conditions when the solute is in liquid form it is more suitable to represent its concentration in volume/volume percent (v/v%). As an example, 12% alcohol means 12 ml of alcohol dissolved in 88 ml of solvent. Weight by volume percentage (% w/v): This method represents the amount of solute as grams and solution in milliliters. An example will be a 5% (w/v) $FeCl_3$ solution where 100 ml of the solution will take 5 g of $FeCl_3$. The following are some equations that are utilized to represent the concentration of solute in the given solvent.

$$\text{Mole} = \frac{mass\ (g)}{gram\ molecular\ weight\ (g)}$$

$$\text{Molarity of a solution} = \frac{number\ of\ mole\ of\ solute}{number\ of\ liters\ of\ solutions}$$

For example, 1 molar solution of H_2SO_4 contains 98.08 g H_2SO_4 in a liter of solution.

$$\text{Molality of a solution} = \frac{number\ of\ mole\ of\ solute}{number\ of\ kilograms\ of\ solvent}$$

A molal solution comprises 1 mol of solute dissolved in 1 kg solvent. Molality is properly expressed as mol/kg.

$$\text{Normality of a solution} = \frac{number\ of\ gram\ equivalents\ of\ solute}{number\ of\ liters\ of\ solutions}$$

$$\text{Gram equivalent weight (as oxidant or reductant)} = \frac{formula\ weight\ (g)}{difference\ in\ oxidation\ state}$$

In the past, mill equivalent (mEq) was utilized to represent the concentration of electrolytes in plasma. Now, the suggested unit for representing the concentration of electrolytes in plasma is mill moles per liter (mmol/L). Many international clinical laboratory organizations and national professional societies have accepted the SI unit in its broad application. Base, derived, and supplemental units are the three classes of SI units. The United States is among the few countries that have yet to accept SI units.

5.2.1.2 OBSERVATION DATABASE

LOINC (logical observation identifier names and codes) and the IFCC/ IUPAC (International Federation of Clinical Chemistry/International Union of Pure and Applied Chemistry) systems are the two systems developed for the expression of clinical laboratory results. LOINC system is a universal coding system for the reportage of laboratory and other clinical interpretations to facilitate the electronic broadcast of laboratory data within and between the institutions (http://www.loinc.org).

5.2.1.3 REAGENTS

The feature of analytical outcome founded by the laboratory is a direct indication of purity of the chemicals used as analytical reagent. The availability and features of the reference materials utilized to standardize the assay and monitoring their analytical performance is very important. Reagent or analytical reagent grade chemical meets ACS (American Chemical Society) specifications. It meets the criteria for chemicals utilized in numerous high-purity applications. These are available in lot- analyzed and maximum impurity reagent forms. The reagent grade chemicals have higher purity, suggested for qualitative or quantitative analyses. Some of the analytical techniques require extremely pure reagents. According to specific requirement manufacturer offer especially purified chemicals. These chemicals are designated as spectrograde, nanograde, HPLC grade and LC-MS grade, as per technique in used. It is recommended that chemicals labeled with, technical, practical, or commercial grade must not be applied in clinical analysis before proper purification.

5.2.1.4 REFERENCE MATERIAL

Primary and secondary reference materials are used in the clinical labora-tory. Primary reference materials are highly purified chemicals. They can be weighed directly to yield a solution of desirable concentration. These chemicals are 99.98% pure, supplied with a credential of analysis for each lot, stable substances and not hygroscopic in nature. On the other hand, solution concentration of secondary reference materials cannot be produced by weighing the solute and dissolving it in an identified amount in a volume of solution. Their concentration is usually determined by the analysis of a portion of solution by an acceptable reference method. In this method, a primary reference material is utilized for calibration.

5.2.1.5 ALTERATION IN SOLUTION CONCENTRATION

In clinical laboratories, sometime solute concentration is needed to increase or decrease in the given solution (clinical sample). Various procedures such as dilution, concentration, and filtration of the solution are utilized for the purpose. When we add the solvent in a solution, the concentration/activity of the solution is reduced and the process is termed as dilution. A serial dilution is a successive set of dilutions in a mathematical order (Figure 5.3). During a dilution, the subsequent equation is utilized to define the volume (V1,) necessary to dilute a certain volume (V2) of a solution of an identified concentration (C1) to the anticipated lesser concentration (C2):

$$C_{1 \times} V1 = C_2 X V$$

$$V_2 \frac{C1 \ X \ V1}{C2}$$

Likewise, the equation might also be used to estimate the concentration of diluted solution when a given volume is added to the starting solution.

Evaporation (conversion of a liquid/volatile solid into vapor) is utilized to eliminate liquid from a sample thereby increase the concentration of analyte(s) left behind. In clinical laboratory, lyophilization or freeze-drying techniques are used to prepare calibrator, control material, reagents, and specific specimens for analysis. In this technique, the material first entails freezing at $-40°C$ or less and then exposed to a high vacuum. The solid mate-rial present in the solution is obtained in a dried state. Passage of a liquid via a filter under the effect of gravity, pressure, or vacuum is known as filtration. This procedure removes particulate matter from the liquid. In the clinical laboratories, mostly filter paper and plastic membrane with controlled pore size are used for filtration. When extremely fine filters are applied to eliminate dissolved particles, the technique is known as ultrafiltration.

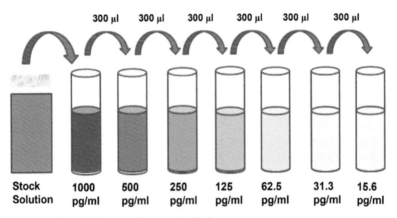

FIGURE 5.3   Serial dilution of the stock solution.

5.2.1.6   *REAGENT GRADE WATER*

Reagent grade water is utilized in clinical laboratory experiments for the preparation of the reagents and solutions. The water brand is suffixed by CLRW (clinical laboratory reagent water) and is "pure" water. Different techniques have been utilized to prepare reagent grade water.

5.2.1.6.1   *Distillation*

One of the key physical properties of a pure liquid is its boiling point. Distillation is applied to purify a mixture of liquids. The liquid needs to be heated up to boiling point temperature followed by the condensation of vapor.

5.2.1.6.2   *Ion Exchange*

The technique is utilized to soften and de-ionize the water. The technique utilizes the electrolyte exchange (fixed to ion-exchange resins) with ions in the water. The water permeates via ion-exchange resins and exchange of ions with electrolyte present in water occurs on the beads. Softening and deionization are the mainly applied ion-exchange approaches used to prepare reagent grade water.

Different grades of water such as type I, II, and III are available for clinical biochemistry experiments. Certain qualitative experiments (such as urine analysis) and washing of glassware are executed by type III grade water. For general laboratory purposes, type II water is used. The water should be free from chemical or bacterial contamination. Clinical laboratory test such as trace elements, electrolyte measurement, and enzyme activity needs type I grade of water. These procedures require minimal interference and maximal precision and accuracy. Type III and II grade water can be stored for further use but the type I grade water should not be stored for further usage. For clinical laboratory practices, reagent grade water should purify once in a week. The water should be tested for microbial contamination, resistivity, pH, and soluble silica. The water should be verified through HPLC (high-performance liquid chromatography) by using a gradient program and ultraviolet detector.

During interpretation of a clinical test result, different physiological factors must be considered because they have an effect on the test results resulting into inappropriate refinement among normal and abnormal result values. Reference range of some analytes (for example, serum creatinine) differ for men and women; depends on age of the patients (neonates, children, adults, and the elderly); affected by diet (sample collected at fasting or after a meal) and time when the sample was taken (during the day or night); patient having stress or anxiety; effect of exercise; medical history; pregnancy and menstrual cycle, etc. All the mentioned biological factors if not considered during the interpretation of the test result will give false diagnosis.

5.2.2 ANALYTICAL TECHNIQUES

5.2.2.1 PHOTOMETRY AND SPECTROPHOTOMETER

Energy transmits as electromagnetic waves of their respective frequency and wavelength. Photometry is the quantification of luminous intensity of a source light directed on a surface. Spectrophotometry determines the intensity of light of a particular wavelength. When an incident light beam with intensity I_0 pass across a square cell comprising a solution of a substance that absorbs light of a specific wavelength, the intensity of the conveyed light beam is less than l_0, and the transmitted light (T) is defined as T= I_s/I_0. Beer's law correlates the substance concentration with light absorbed/transmitted by the substance. The law states that the concentration of a compound is directly proportional to the amount of light absorbed or inversely proportional to the logarithm of the conveyed light. Mathematically, Beer's law can be given as:

$$A = abc$$

where, A = Absorbance; a = Proportionality constant given as absorptivity; b =Light path in centimeters; c = Concentration of the absorbing compound, generally given in grams per liter.

Spectrophotometers are classified as single or double-beam. The elementary constituent of a spectrophotometer include, a *light 444 source* (incandescent lamps and lasers); *monochromator* (a to isolate light of the desired wavelength); *fiber optics* (better directional control of the light beam within the geometrical confines of an instrument); *cuvettes* (a tiny container utilize to keep liquid samples are going to be analyze in the light path of spectrometer); *a photodetector* (changes light into electric signal proportional to the quantity of photons striking on its photosensitive surface); a readout device (displays electrical energy from a detector); a recorder and a computer (Karnik et al., 2006).

5.2.2.2   ELECTROCHEMISTRY

Several analytical approaches utilized in the clinical laboratories apply electrochemical measurement such as potentiometry, voltammetry/ amperometry, conductometry, and coulometry. Potentiometry refers to the measurement of alteration in electrical potential amongst two electrodes (half-cells) inside an electrochemical cell. Potentiometric sensors are extensively applied clinically to measure pH, PCO_2 and electrolytes (Na, K, Cl⁻, Ca^{2+}, Mg^+, and Li) present in whole blood, plasma, urine, and serum and as transducers for developing biosensor regarding metabolites of clinical importance. Voltammetric and amperometric techniques are generally utilized electroanalytical methods in the clinical laboratory. In this technique an external voltage is applied to a polarizable working electrode (measured versus a suitable reference electrode: $E_{appli} = E_{work} - E_{ref}$), and the subsequent anodic (for analytical oxidations) or cathodic (for analytical reductions) current of the cell is observed. The current is proportional to the analyte concentration residing in the test sample (Karnik et al., 2006).

5.2.2.3 ELECTROPHORESIS

The technique deals with the motion of charged solute/particle inside a liquid medium, in an electric field. There is various kind of electrophoresis, but zone electrophoresis is most frequently utilized for clinical applications. In this technique, the sample is dissolved in a buffer and the charged molecules move as zones, typically inside a porous supportive medium, for example, agarose gel film. The process generates zones of protein on the gel film, each carefully differentiated from neighboring zone. The gel is stained with a protein-specific dye for the visualization of zones. The medium is dried and the zones are quantified in a densitometer.

5.2.2.4 CHROMATOGRAPHY

The technique is applied in the clinical laboratories to differentiate and enumerate various kind of analyte (Figure 5.4). Chromatography is a method which deals with the separation of desired solute in a given mixture due to its differential distribution among stationary and mobile phases. Stationary and mobile phases are used in this technique. The mobile phase passes the sample via a column, bed, or layer (stationary phase). As the mobile phase (having solute/analyte) pass the stationary phase some solute particles react with the stationary phase and their migration rate is decreased. Some solute particles have less affinity with the stationary phase and reside in the mobile phase. They migrate at a faster rate. Consequently, the lower affinity solute differentiates from solute with higher affinities to the stationary phase. Different chromatographic separation mechanism such as ion-exchange, partition, adsorption, size-exclusion, and affinity are known to separate the solute. Mostly clinical methods apply chromatographic differentiation depending on ion-exchange and partition mechanisms. Chromatography based on the ion-exchange separation mechanism is known as ion-exchange chromatography. This technique depends on an interchange of ions between a charged stationary surface and ions of the opposite charge of the mobile phase. Partition chromatography (PC) deals with the differential scattering solute particles among two immiscible liquids. Separation depends on alterations in the comparative solubility of solute molecules among the mobile and stationary phases. PC may be either gas-liquid chromatography or liquid-liquid chromatography.

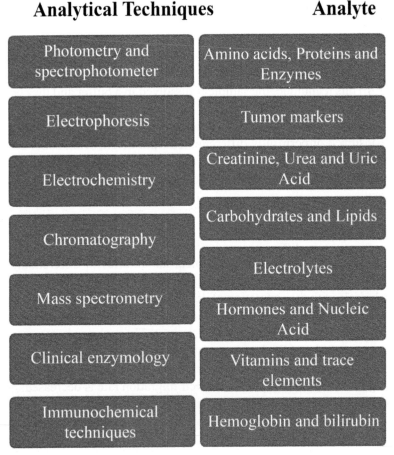

FIGURE 5.4   Different analytical techniques and analyte utilized in clinical biochemistry laboratories.

5.2.2.5   MASS SPECTROMETRY (MS)

In mass spectrophotometry, the target molecule first gets ionized followed by separation and measurement of the mass of molecule and its fragment. Recognition of the combined ionic species is based on mass-to-charge (m/z) ratio of ion. The respective profusion of every ion plotted as a function of its m/z ratio generates a mass spectrum. The fragmentation of ions at particular bonds resides on their chemical nature. It is possible to determine the structure of an analyte from its mass spectrum. Mass spectrometry (MS) is a quantitative and qualitative analytical technique, utilized for the measurement of an extensive range of clinically relevant analytes.

5.2.2.6 CLINICAL ENZYMOLOGY

Catalyst, a substance which raises the rate of chemical reactions but does not get consumed or permanently altered. Enzymes are known as a biological catalyst. All enzyme molecules possess the primary, secondary, and tertiary structural characteristics of protein. The catalytic ability of an enzyme molecule is generally based on the integrity of its structure (Karnik et al., 2006). Different enzyme alternatives might occur inside a single organ or within a single kind of cell and are termed as isoenzymes. The activity of isoenzymes is different at the organ, cellular, and subcellular levels. It is largely acknowledged that in pathological conditions the activity of specific isoenzyme gets altered. In the clinical laboratory, the activity, or protein mass of enzyme in plasma or serum are measured. Mostly enzymes are intracellular and normally present at small concentrations in the serum. In diseased condition, cells of a respective organ/tissue get damaged and the intracellular enzyme leaked into blood. Measurement of enzyme concentration in individual sample during the disease, it is possible to conclude the location and feature of pathological changes in body tissue. Various kind of analytical methods are employed to measure isoenzyme activity such as chemical inactivation, chromatography, electrophoresis, isoelectric focusing, and differences in catalytic property.

5.2.2.7 IMMUNOCHEMICAL TECHNIQUES

Antibodies have ability to bind precisely with extensive range of antigens. The antigen might be natural or synthetic; carbohydrate, nucleic acid, protein, lipid, or other molecule. Immunochemical techniques deal with different immunochemical reactions of immunoassays. The immunoassays are very sensitive and specific. In a clinical immunoassay, an antibody is utilized as a reagent to perceive the analyte of interest. In immunoassay, the analyte is an antigen. Different immunoassay deal with fine specificity and high affinity of antibody towards specific antigen and also the unique ability of antibody to cross-link the antigens. These two properties of antibodies allow us to identify and quantify analyte by a variety of methods. Different qualitative (passive gel diffusion, immunoelectrophoresis, western blotting) and quantitative (electroimmunoassays, radial diffusion, nephelometric, and turbidimetric assays, labeled immunochemical assays) immunochemical techniques are used in clinical laboratory experiment. Immunoelectrophoresis is applied to differentiate and identify the protein in a clinical sample such as serum or spinal fluid. Enzyme immunoassay utilizes the catalytic feature of enzyme for the perception and quantification of immunological reactions. Glucose-6-dehydrogenase, ALP, and horseradish peroxidase are the common enzymes in an enzymatic immunoassay.

5.2.3 ANALYTE

5.2.3.1 AMINO ACIDS AND PROTEINS

Amino acids are building block of protein. DNA has codes that decide the sequence in which amino acids are aligned in a protein. The special arrangement of amino acids for a certain protein is necessary for its proper function. Mutation in nucleic acid might result in wrong amino acid incorporation for a given protein. The wrong organization of amino acids in a protein will hamper its function. The primary repertoire of amino acids for protein synthesis in the body is diet. Although some amino acid are synthesized *in vivo*. Measurement of amino acid in physiological fluids is an important way for the conclusion of many pathological conditions. Amino acids could be measured quantitatively by numerous techniques like capillary electrophoresis, gas, and ion-exchange liquid chromatography, HPLC, and tandem MS is the given biological fluid.

5.2.3.2 ENZYMES

Enzymes may be extracellular or intracellular. Some enzyme is tissue-specific and some are ubiquitous in the body. In blood, some amount of this enzyme can also be found in normal condition. In disease condition, the tissue/organ of the body get affected and the cells damage. Breaking of cell membrane allows the intracellular enzyme leakage into the blood. Measurement of enzyme concentration in an individual's blood makes the basis of different clinical prognosis and diagnosis. For example, elevated levels of CK (serum creatine kinase activity) in muscular dystrophy are a clinical test. Techniques such as PAGE (polyacrylamide gel electrophoresis) are utilized for the assay.

5.2.3.3   TUMOR MARKERS

The tumor is an unidentified growth of abnormal tissue that is often uncontrolled and progressive. The tumor may form at any site of the body. Based on that cancer might be of prostate, breast, liver, lung, etc. Tumor produces some substances that differentiate them from normal tissue. These substances are referred to as tumor markers and generally are secreted in the biological fluids (mostly blood). Tumor markers are cancer-specific and nonspecific both. Mostly tumor marker's blood concentration indicates the activity and volume of tumor. For early diagnosis/screening the tumor marker need to be clinically precise (for a given cancer) and sensitive (for detection in small quantity). Different groups such as NACB (National Academy of Clinical Biochemistry) and EGTM (European group on tumor markers) released the procedures for the selection and use of tumor markers in the clinic. Enzyme assay, immunoassay, receptor assay, chromatography, and electrophoresis are techniques utilized for the measurement of tumor marker measurement.

5.2.3.4   CREATININE, URIC ACID, AND UREA

Creatinine is the end result of phosphocreatine disintegration. Urea is released by the nitrogen backbone of amino acids after metabolism and responsible for more than 75% of the nonprotein nitrogen, ultimately excreted. Uric acid is chemically 2, 6, 8-trihydroxypurine found in low amounts in mammalian urine. It is the major outcome of the metabolism of purines in humans. Creatinine, uric acid, and urea are nonprotein nitrogenous metabolites which are unfurnished from kidney after glomerular filtration. Measurements of metabolite concentrations of metabolites in plasma/serum are used as kidney function indicators.

5.2.3.5   CARBOHYDRATES

Carbohydrates are extensively distributed in animal and plants. They constitute nucleic acid and provide a source of energy. Excess of carbohydrate is stored as fat and glycogen in adipose tissue and liver/muscles. Under many physiological circumstances (feeding, fasting, or severe exercise) insulin, epinephrine, glucagon, and hormones maintain the concentration of glucose in the blood. Diabetes mellitus is the most widely occurring disorder of abnormal carbohydrate metabolism. Glucose measurement is the generally performed clinical test in hospital and other healthcare laboratories.

5.2.3.6   ELECTROLYTES

Water is the major solvent in a living organism. Its homeostasis is crucial to all life. The four major electrolytes: sodium, potassium, chloride, and bicarbonate play important role in the maintenance of osmotic pressure and water balance in different body fluid compartment. They also maintain pH, proper heart, and muscle function, oxidation-reduction reactions, and act as cofactor for enzyme. Alteration in electrolyte concentration beyond normal values might be the cause or the consequence of much disorder. That's why the electrolyte concentration determination is main the routine work in a clinical laboratory.

5.2.3.7   HORMONES

The hormone is a chemical moiety produced by cells and has a definite regulatory effect on body physiology. They are autocrine and paracrine in nature. Generally, hormones act through receptors present on the target cell. Chemically hormones are steroid, amino acid derivatives, or polypeptide/protein. Hypo (less than normal) or hyper (more than normal) secretion of hormone leads to various diseases and disorders. It is measured by a variety of techniques/assays such as receptor assay, immunoassay, bioassay, MS, and gas or liquid chromatography.

5.2.3.8 *VITAMINS AND TRACE ELEMENTS*

Vitamins can be defined as organic compounds compulsory in the food for growth, health, and reproduction. Trace elements are the elements which are needed in a very low quantity for to proper functioning of body. Originally, trace element term was applied to define the remaining amount of inorganic analyte that has been quantitatively determined in the sample. Vitamins and trace elements insufficient amount are a necessity for health and development of the body. United States made a Suggested (RDA: recommended dietary allowances) dietary allowance for vitamins and trace elements (Burtis and Bruns, 2014). Now a day's many sensitive analytical methods are present to determine the inorganic micronutrients present in very small concentrations in body fluids and tissue more accurately.

5.2.3.9 *HEMOGLOBIN AND BILIRUBIN*

Hemoglobin is the pigment responsible for conveyance of oxygen to the cell and carbon dioxide from the cells to the lung. Iron of the heme group is the core of hemoglobin function. Hemoglobin has a limited lifespan after which it degrades into a waste product known as bilirubin. Hemoglobin, iron, and bilirubin are analytes as their measurement in biological fluid gives an idea about iron's homeostasis in the body. Iron homeostasis is a major requirement for the appropriate functioning of body. Any alterations in it might leads to many diseases and disorder. Iron deficiency and overload both are the major disorders resulting from irregular iron metabolism. Serum iron and ferritin, iron- binding capacity and transferrin saturation are some common methods applied to measure iron and respective analytes. Cyanmethemoglobin technique is generally used for the measurement of hemoglobin concentration in venous or capillary blood. Besides the above-mentioned analyte, some other includes nucleic acid, lipids, therapeutic drugs, and toxic metals which have their own importance in diagnosis of various clinical complications.

KEYWORDS

- clinical laboratory reagent water
- International system of units
- logical observation identifier names and codes
- mass spectrometry
- partition chromatography
- recommended dietary allowances

REFERENCES

Burtis, C. A., & Bruns, D. E., (2014). *Tietz Fundamentals of Clinical Chemistry and Molecular Diagnostics-E-Book*. Elsevier health sciences.

Karnik, R., Castelino, K., & Majumdar, A., (2006). Field-effect control of protein transport in a nanofluidic transistor circuit. *Appl. Phys. Lett.*, *88*(12), 123114(1–3).

CHAPTER 6

Biochemical Aids to Clinical Diagnosis

PREM PRAKASH KUSHWAHA,[1] P. SESHU VARDHAN,[2]
K. VISHNUPRIYA,[3] SHARMISTHA SINGH,[4] and SHASHANK KUMAR[1]

[1]*School of Basic and Applied Sciences, Department of Biochemistry, Central University of Punjab, Bathinda, Punjab–151001, India, Tel.: +91 9335647413,*
E-mail: shashankbiochemau@gmail.com (S. Kumar)

[2]*School of Biotechnology, Jawaharlal Nehru Technological University, Kakinada–500085, Telangana, India*

[3]*Protein Bioinformatics Lab, Department of Biotechnology, Indian Institute of Technology, Madras – 600036, Tamil Nadu, India*

[4]*Department of Biochemistry, University of Allahabad, Allahabad, India*

ABSTRACT

Microorganism is typically responsible for the several human diseases. This disease can be identified by various assays using different biological sample. Numerous enzymes and their levels present in the different biological sample reveals the condition of the patients. Substances present in the serum such as urea and creatinine explores the condition of the kidney health. Enzymes, ions, and hormones like alkaline phosphatase (ALP), lactate dehydrogenase (LD), neuron-specific enolase, cathepsins, calcitonin, calcium, phosphorus, blood urea nitrogen (BUN) and electrolytes demonstrates several organs function and their health.

6.1 INTRODUCTION: CARDIOVASCULAR DISEASES (CV)

Cardiovascular disease (CV) is the major cause of enervation as well as hasty death worldwide. Atherosclerosis is an underlined pathology of CV;

require an extensive time to grow. One can have its symptoms generally in middle age. In CV, sudden and frequent occurrence of acute coronary proceedings occurs and is often fatal. CV might diagnose by different laboratory tests and imaging techniques. Some of the common tests for CV diagnosis are as follows:

- Tests for heart disease risk factors include cholesterol, lipid (LDL, HDL, Triglycerides) and fats levels in the blood.
- Glycosylated Hb in blood is used to diagnose diabetes. For inflammation (that might lead to heart disease), different protein tests such as C-reactive protein (CRP) test, and apolipoprotein A1/B are used.
- During heart attack, myocardial cells release some protein and factors into the bloodstream. These factors are important signature of a recent heart attack. Cardiac Troponin-T, increased homocysteine level and asymmetric dimethylarginine, brain natriuretic peptide (BNP), fibrinogen, and PAI-1, are some important marker of heart attack.

6.2   FUNCTION AND DISEASES RELATED TO KIDNEY

In our body evacuation of water-soluble waste products is the key role of kidney. Besides filtration and secretion is also the major responsibility of kidney. Deregulation of these functions results in the decreased excretion and accumulation of waste products in the body. Assessment of various proteins and molecules in urine might be used for kidney function:

- Urine examination;
- Urea clearance;
- Serum urea;
- Serum electrolyte levels;
- Serum creatinine;
- Blood urea nitrogen (BUN);
- Calcium;
- Concentration-dilution test;
- Protein;
- Albumin;
- Creatinine clearance;
- Phosphorus;
- Inulin clearance.

6.2.1   URINE EXAMINATION

A qualitative examination of urine provides an idea about nature and site of damage in the renal system. Color, odor, quantity, specific gravity of the urine is some parameter for qualitative assessment of urine. Pus cells, RBC casts, and crystals are some microscopic inspection of the urine sample.

6.2.2   SERUM UREA

Amino acid metabolism and protein catabolism pathways are related to the production of urea in the body. Urea production occurs in the liver by urea cycle which later on undergoes filtration in glomerulus secretion and tubular level re-absorption. Enhancement in serum urea levels indicates renal/ glomerular dysfunction. The normal concentration of serum urea is 20–45 mg/dl and is affected by diet and some non-kidney associated ailment also.

6.2.3   BLOOD UREA NITROGEN (BUN)

BUN generally expressed in terms of serum urea. Urea is a 60 mw molecule has two nitrogen atoms and has collective atomic weight of 28. Thus, in serum, nitrogen contribution to the total weight of urea is 28/60 and corresponds to 0.47. Serum urea level can be simply transformed into a BUN by multiplying it level by 0.47.

6.2.4   CALCIUM

Calcium test for the renal disturbance measures Ca in blood only, not in our bones. This test gives an idea about parathyroid glands, kidney, cancer, and bone problems. The standard range of Ca^{2+} in blood is 8.5 to 10.2 mg/dl.

6.2.5   PHOSPHORUS

Phosphorus is an important constituent of bones and teeth, support nerve function and muscle contraction. Extra quantity of phosphate in the blood is decanted by the kidneys and flow away in the urine. Elevated levels of phosphorus have been in the patients of severe kidney disease.

6.2.6   PROTEIN

Protein level estimation in urine is a very sensitive and wide-range screening test for renal ailment. It is strikingly augmented in renal disease. The highest degree of proteinuria was found in nephrotic syndrome (>3–4 g/day). The renal disease having nephrotic syndrome, urinary protein flow rate is typically about 1–2 g/day.

6.2.7   SERUM CREATININE

Creatine (Cr) is a trivial tripeptide mostly occurs in muscles and is nontoxic. It liberates from the muscles and gets converted into Cr. Cr is not a noxious product and is applied as a symbol of renal function. Normal serum Cr level is 0.6–1.5 mg/dl. It is a preferable pointer of renal and glomerular function. Cr is mostly calorimetrically measured by Jaffe's method in clinical laboratories.

6.2.8   UREA CLEARANCE

Urea clearance stands for the excretion of blood urea with the help of kidney within one minute. Urea clearance measured by urea level in the blood, urine, and quantity of urine expelled over a break of one hour.

6.2.9   CREATININE CLEARANCE RATE

Creatinine filtration at glomerulus and reabsorption at tubular level is irrelevant. Due to this creatinine, clearance may be used to quantify glomerular filtration rate (GFR).

6.2.10   INULIN CLEARANCE

Inulin is a trivial polysaccharide made up of fructose and is a very low molecular weight compound. It is utilized for the measurement of GFR.

GFR is the quantity of blood that passes and sieved through glomerulus per minute.

6.2.11 CONCENTRATION TEST

Kidney has ability to concentrate lodged water and urine by increasing its reabsorption from the glomerular filtrate at the tubular level. This function of kidney is a measure of tubular function. The conspicuous gravity of as minimum as one sample should be 1.025 or above, in a normal person. Specific gravity below the 1.025 is an indication of tubular dysfunction.

6.2.12 ELECTROLYTES

Kidney balances the water content and its excretion which leads into the maintenance of electrolyte balance in the body. It actively reabsorb/excrete electrolytes and thereby maintains the electrolyte balance.

Filtration and reabsorption of electrolytes occurred at the glomerulus and tubular level respectively. Problem at tubular level results in non-absorptive and excessive loss of electrolytes in the urine. Important serum electrolytes to explore the tubular level problem are sodium (135–145 mmol/liter), potassium (3.5–5 mmol/liter), and chloride ions (95–105 mmol/liter).

6.3 LIVER DISEASE

The adult liver weighs approximately 1.2–1.5 kg, positioned underneath the diaphragm in the right upper part of abdomen, protected by the ribs. It carries out a number of excretory, synthetic, and metabolic functions. Endogenous and exogenous organic anions are removed from sinusoidal blood, bio-transformed, and expelled into the bile or urine. Excretory function evaluation delivers appreciated clinical information. Measurement of plasma concentrations of endogenous products such as bilirubin and bile acids, and the assessment of clearance rate of exogenous products (aminopyrine, lidocaine, and caffeine) are the most recurrently used examinations for the diagnosis of liver delinquent. A number of circumstances are symptomatic of liver problem such as jaundice, portal hypertension, obnoxious hemostasis, and enzyme secretion into body fluids. Jaundice is characterized by the appearance of mucous membrane, yellowish skin, and bilirubin deposition (sclera).

Jaundice is frequently ostensible when plasma bilirubin concentration reaches up to 34 to 51 μmol/L (2–3 mg/dL). Ascites is the outpouring and gathering of fluid inside the abdominal cavity. It is uncomfortable and compromise respiration but not a life-threatening state. Portal hypertension is an increase in the difference between plasma and ascitic fluid albumin concentration. A gradient > 1.1 g/dL is symptomatic of ascites triggered by portal hypertension. Ascites predisposes to spontaneous bacterial peritonitis is defined as bacteremia (i.e., presence of bacteria in the blood) in the absence of power-driven disruption of bowel. The diagnosis can be achieved by ascitic fluid examination; >250 neutrophils per microliter, or >500 in the death of a positive blood culture, is considered as positive diagnostic. Sometimes patients having liver disease showed normal hepatic function tests. In liver disease increased levels of plasma, activity of many cytosolic, mitochondrial, as well as membrane- associated enzymes has been reported.

Several factors administrate the potential of liver enzymes capability to support diagnosis, including their tissue specificity, subcellular distribution, comparative degree of enzyme commotion in liver and plasma, pattern of release, and clearance from the plasma. Although jaundice is a crucial clinical event tending leading to recognition of acute hepatitis, is often absent. AST activity more than 200 IU/L, or ALT activity more than 300 IU/L, has clinical specificity and sensitivity of more than 90% for acute hepatitis (Burtis and Bruns, 2014).

6.4   THYROID DISORDER

Thyroid gland has butterfly-like appearance, situated in the opposite of neck impartial upstairs the trachea in mature human being. Thyroid gland of a mature human weighs about 15–20 g and comprise two lobes linked by isthmus. Thyroid follicles are the secretory unit of thyroid gland. Epithelial cells form an exterior layer on individual follicle encompassing an amorphous material called colloid. Colloid is principally composed of thyroglobulin (Tg) has trivial extents of iodinated thyroid-albumin. The thyroid gland similarly has additional type of cell known as parafollicular or C cells. These cells produce the polypeptide hormone called calcitonin. These cells are restricted within the follicular basement lamina or occur in clusters in the inter-follicular area. The two hormones secreted by thyroid gland (thyroxine and triiodothyronine) are commonly known as T4 and T3 respectively. Thyroid hormones have much important biological effect. A major function of these hormones is to regulate BMR and calorigenesis via augmented oxygen feasting in tissue through the possessions of thyroid hormone on membrane transportation (cycling of Nai/K^+-adenosine triphosphatase (ATPase) with increased synthesis and consumption of adenosine triphosphate) and enhanced mitochondrial metabolism (stimulation of mitochondrial respiration and also oxidative phosphorylation). Almost all laboratory tests for thyroid function are commercially available in kit form for on automated immunoassay instruments. High sensitivity assays for TSH are available for various signal detection such as chemiluminescence and assays with low end detection limits in the range of 0.01 to 0.05 mIU/L. Secretion of TSH occurs in a circadian fashion: highest concentrations prevail at night (between 200–400) and lowest concentrations occur in the range of 1700–1800. Low amplitude oscillations also occur throughout the day. Increase in TSH in vespertine is lost in critical illness and after surgery. TSH flows immediately after birth, peaking at 30 minutes at 25 to 160 mIU/L; then values drop down to cord blood concentration in three days and reach at adult values in the first weeks of life. Immunoassays of total T4 measures both free and protein-bound thyroxine. Accurate measurement of total endogenous hormone therefore requires alienation of T4 from its serum transport proteins because 99.97% of the T4 circulates tightly bound to albumin, TBG, and TBPA. Hypothyroidism and hyperthyroidism are the two principal pathological circumstances related to thyroid gland. Patients with hyperthyroidism characteristically have serum TSH concentrations <0.05 mIU/L (Burtis and Bruns, 2014).

6.5 PITUITARY DISORDER

The adrenal cortex of human (outermost layer of adrenal gland) secretes three major steroid hormones possess extensive of physiological role. These include mineral ocorticoids, glucocorticoids, and adrenal androgens. Steroid hormones are steroidal in nature and act as hormone. Cholesterol is the precursor of most of the human steroid hormones produced primarily in adrenal glands and gonads. The liver is the major site of steroid metabolism. The kidney and gastrointestinal tract, however, also carry out vital metabolic transformation of steroids. Cortisol, a glucocorticoid formed from cholesterol in zona fasciculata and reticular is of human adrenal cortex. Mineralocorticoids control salt homeostasis (Na preservation and K loss) and extracellular fluid volume. Aldosterone is the greatest effective and naturally occurring mineralocorticoid produced entirely in the zona glomerulosa section of the adrenal cortex. This zone contains characteristic aldosterone synthase, an obligatory enzyme in the aldosterone synthetic pathway. The hormones are generally produced in a definite amount in the body. Immunoassay is the furthermost extensively used method for the measurement of cortisol, aldosterone, and DHEA. Measurement of cortisol concentrations and basal ACTH along with ACTH stimulation test is suggested to diagnose primary adrenal insufficiency. Basal plasma ACTH concentrations $>I_{50}$ microgram/mL with serum cortisol concentrations <10 microgram/dL are the diagnostics of adrenal insufficiency. Cushing syndrome is the outcome of autonomous, extreme cortisol production leading to the classic symptoms. Different simple screening tests are available for the detection of Cushing syndrome such as measurement of 24-hour urinary free cortisol. Under normal circumstances<2% concentration of secreted cortisol is present in urine in the form of free cortisol. In general, 24 hours urinary free cortisol concentration <100 micrograms/day excludes the diagnosis of Cushing syndrome, and concentrations >120 suggest the occurrence of Cushing syndrome. The clinical diagnostic accuracy is more than 90% subjected the test was properly performed. An elevated excretion rate results in the overproduction of cortisol. However, inappropriate timing of the urine specimen, commensurate use of a diuretic, high salt consumption, depression, and stress has been detected to originate false-positive test results. Urine cortisol measurements do not underlay the diagnosis and an abnormal result should be followed by repeated or provocative testing.

6.6 TUMOR MARKERS

A tumor marker is a tenor originating from tumor or by the host in repercussion to a tumor. Tenor helps in the tumor differentiation from ordinary tissue or regulates the occurrence of a tumor depending on the measurements in blood or secretory products. Its availability in tissue, cells, or body fluids and gauged can be monitored qualitatively or quantitatively by immunological, chemical, or molecular diagnostic techniques. Tumor marker varies from tumor to tumor. Some of the tumor markers are expressed in all type of cancer. However, some of the tumor markers precisely expressed in particular cancer. In cancer diagnosis, these nonspecific tumor markers are not qualified. In several cases, tumor activity and volume are regulated by blood concentrations of tumor markers. For proper diagnosis of a tumor, the marker has to be precise with respect to tumor type and should be sensitive to spot small tumor or early-stage screening. Inappropriately, some markers are precise only for a single individual type of tumor called, tumor-specific markers. Sometimes markers are associated with dissimilar tumors of a similar tissue type and are known as called tumor-associated marker. Tumor marker showed high expression in cancer tissue/blood in patients, in comparison to benign tumor/blood of normal individual in normal. Tumor marker help in the assessment of cancer progression status after the preliminary treatment. Most of the tumor marker is an enzyme, hormone, oncofetal antigen, carbohydrate, blood group antigen, protein, receptor, or genes.

Bence Jones protein was the first tumor marker documented in 1987. As we lead off the 21st century, new technologies are being applied for the discovery of tumor markers and their clinical application. Notable among these discoveries are the introduction of genomics and proteomics approaches such as measurement of complementary DNA (cDNA), protein, and tissue microarray, and mass spectrometry (MS) usages as a diagnostic tool. Furthermore, the advent of bioinformatics techniques, including neural networks, logistic regression, support vector machine, and other algorithms, are facilitating the utilization of multiparametric (multiple analytes) analysis of cancer prognosis, diagnosis, and the prediction of therapy.

In general, tumor markers are used as a diagnosis and monitoring tool to check the therapeutic effects. Ideally, tumor marker should be synthesized by cancer cells and in measurable amount in the body fluids. It should not exist in healthy person or in benign circumstances. Therefore, it can be used for screening for cancer occurrence in asymptomatic person of general population. Numerous tumor markers exist in normal, benign, and cancer tissues and are not precisely adequate to be used for cancer screening. However, if the frequency of cancer is huge among the certain populations, screening could be feasible. The staging of clinical cancer is assisted by marker quantification (i.e., the serum concentration of some markers reflect tumor burden). At the time of the diagnostic, marker value could be utilized as a prognostic indicator for cancer progression and patient survival. This is conceivable for an individual patient, but dissimilar markers concentration originate by diverse types of tumors do not typically allow one to control the tumor prognosis from the preliminary concentration. However, after initial treatment like surgery, the quantity of marker should be decreased. Reducing rate of the marker can be detected by measuring its half-life. If the post- treatment half-life is longer than the expected half-life, the treatment has not been considered as positive to eradicate the tumor. Reduction in marker magnitude might reflect the success rate of the treatment (Burtis and Bruns, 2014).

Enzymes are present at abundant concentrations in the cell. Their secretion inside the systemic circulation results in tumor necrosis or change the cell membrane permeability of the tumor. Augmented enzyme activities are also detected in pancreatic or biliary duct obstructions and in renal insufficiency. The degree of enzyme release may also be determined by the intracellular occurrence of the enzyme. By the time enzyme secreted into the circulatory system, tumors metastasis might have occurred. Mostly enzymes are not unique for a specific organ. Therefore, enzymes are furthermost appropriate as nonspecific tumor markers; however, elevated enzyme activities may indicate the presence of malignancy. Isoenzymes and numerous forms of enzymes may provide additional organ specificity.

6.6.1   ALKALINE PHOSPHATASE (ALP)

ALP originates in liver, bone, or placenta; however, the ALP in sera of normal adult originates primarily in liver or biliary tract. In both types of liver cancers i.e., primary, and secondary, elevated activities of ALP are seen. The serum ALP activity is used for the assessment of liver metastases and showed better correlation with the degree of liver involvement in metastasis than its other function tests.

6.6.2   LACTATE DEHYDROGENASE (LD)

LD is an enzyme involved in glycolytic pathway and is released as an outcome of cell damage. Elevated levels of LD in malignancy are rather nonspecific. It has been confirmed in various cancer including liver, acute leukemia, non-Hodgkin lymphoma, nonseminomatous germ cell testicular cancer, neuroblastoma, seminoma, and other carcinomas of colon, breast, stomach, and lung. The serum LD levels have been revealed to associate with tumor mass in solid tumors and deliver a prognostic gage for disease progression.

6.6.3 NEURON SPECIFIC ENOLASE

Enolase is another glycolytic enzyme commonly known as phosphopy- ruvatehydratase. In cells of the diffuse neuroendocrine system, neuronal tissues, amine precursor uptake, and decarboxylation (APUD) tissue enolase is in form of neuron-specific enolase (NSE). NSE is originated in tumor of neuroendocrine origin such as small cell lung cancer (SCLC), neuroblastoma, carcinoid, medullary thyroid cancer, pheochromocytoma, melanoma, and pancreatic endocrine tumors.

6.6.4 PROSTATE-SPECIFIC ANTIGEN (PSA)

Prostate-specific antigen (PSA) is a currently available promising tumor marker. It is an organ- specific tumor marker. Prostate cancer (PCa) is foremost cancer in older men, early detection (organ- confined), possibly treatable by radical prostatectomy. Early detection is important for men with a life anticipation of at least ten years.

6.6.5 CATHEPSINS

Cathepsins are lysosomal proteases, its different forms such as cathepsin B, D, and L have been investigated for their role in tumor development and progression. Like other proteases, cathepsins are produced as high molecular weight precursors and require processing for the activation. Thiol- dependent protease cathepsin B (CB) is normally found in lysosomes. It is activated by cathepsin D (CD) and matrix MMPs. Activated CB in turn, activates uPA, and specific MMPs. Cathepsin L (CL) is similar in specificity to that of CB. Cathepsin D, and B, are lysosomal protease; however, CD is an aspartyl group proteases. The manifestation and localization of CB is altered in tumors in comparison to the normal tissue. In tumor tissue, CB would attach the cell membrane or found in secreted form. Increased expression of CB has been found in breast, colorectal, gastric, lung, and prostate tumors, carcinoma, glioma, melanoma, and osteoclastoma, suggesting a link with tumor development and progression. Altered localization of CB has also been found in various tumor tissues, such as colon, thyroid, gliomas, and breast epithelial tumor. The altered manifestation and localization of CB is supposed to be involved in tissue invasion through ECM degradation and growth promotion.

6.6.6   MATRIX METALLOPROTEINASES

MMPs are a group of structurally associated zinc-dependent endopeptidases. They have the ability to degrade the ECM (extracellular matrix) components. Most MMPs are secreted as zymogen, and activated after the removal of 10 kDa amino-terminal domain. In the active form, their proteolytic action is inhibited by specific tissue inhibitors termed as tissue inhibitors of metalloproteinases (TIMPs). The MMPs are functionally grouped into four subgroups based on their ECM specificity: collagenases, gelatinases, stromelysins, and membrane MMPs. MMPs are allied in tissue remodeling and wound repair; however, they are also associated with tumor growth, metastasis, and invasion. Increased expression of MMPs has been associated with tumor aggressiveness as poor prognosis.

6.6.7   HORMONES

Hormones have been documented as tumor marker from more than 50 years. The introduction of precise RIA procedures for a specific hormone has very little cross-reactivity with similar hormones. This quality of hormone made it presumable for the diagnosis of cancer patients. In cancer, secretion of hormones involves two separate routes. First, the endocrine tissue usually secretes additional quantity of hormone. Second, a hormone may be secreted at a distant site by a nonendocrine tissue that usually does not produce hormone. The latter condition is called ectopic syndrome. For example, adrenocorticotropic hormone (ACTH) production by pituitary and small cell of lung are normocopic and ectopic respectively. Consequently, elevation in the levels of a given hormone is not a precise diagnostic for a tumor because a hormone might be produced in a variety of cancers.

6.6.8   CALCITONIN

Calcitonin is a polypeptide of 32 amino acids, with a MW of 3400. It is secreted by thyroid C cells. Normally, it is secreted in response to increased serum calcium levels. It inhibits the secretion of calcium from bone and thus lowers the serum calcium concentration. The serum half-life is about 12 minutes. The serum concentration in healthy individuals is less than 0.1 micro g/L, and an elevated concentration is usually associated with the medullary carcinoma of thyroid. Calcitonin is useful in the detection of familial medullary carcinoma of thyroid (FMTC), an autosomal dominant disorder. Asymptomatic family members of the affected patient get benefit from the screening with computed tomography because basal concentrations of calcitonin are increased in such people.

KEYWORDS

- adrenocorticotropic hormone
- alkaline phosphatase
- amine precursor uptake and decarboxylation
- blood urea nitrogen
- brain natriuretic peptide
- complementary DNA

REFERENCES

Burtis, C. A., & Bruns, D. E., (2014). *Tietz Fundamentals of Clinical Chemistry and Molecular Diagnostics-E-Book*. Elsevier Health Sciences.

CHAPTER 7

Clinical Measurements and Quality Control

PREM PRAKASH KUSHWAHA,[1] P. SESHU VARDHAN,[2]
M. S. SANDEEP VEDANARAYANA,[2] and SHASHANK KUMAR[1]

[1]*School of Basic and Applied Sciences, Department of Biochemistry, Central University of Punjab, Bathinda, Punjab–151001, India, Tel.: +91 9335647413, E-mail: shashankbiochemau@gmail.com (S. Kumar)*

[2]*School of Biotechnology, Jawaharlal Nehru Technological University, Kakinada–500085, Telangana, India*

ABSTRACT

Measurements cover all the required parameters for the clinical and research purpose. Low-quality measurements affect a steady prognosis and diagnosis. Wrong measurements affect epidemiological studies and randomized trials. Statistical analysis deals with reference values, spreading of individual groups, outliers, and reference limits. Administration of entire testing such as test usage and practice guidelines, patient identification, turnaround time (TAT), laboratory logs, transcription errors, patient preparation, specimen collection, transport, and division and spreading of aliquots should be performed after and before analysis.

7.1 INTRODUCTION

Quality measurement is a routine part of the clinical research laboratory to improve the reliability of diagnosis and prognosis. Inaccurate measurements might impede the handling of clinical studies such as randomized trials and epidemiological analyses. Progress in clinical measurement is proportional to new technology or techniques (Burtis and Bruns, 2014).

7.2   ANALYTICAL PROCEDURES AND QUALITY CONTROL

The reference value is an important component of clinical laboratory experiments. The essential components are specifications concerned analysis method (data related to equipment, reagents, calibrators, raw data, and method of calculation) quality control and reliability criteria.

7.3   *STATISTICAL TREATMENT OF REFERENCE VALUES*

After the analysis of reference specimens, the reference values are subjected to a statistical treatment. The treatment includes the segregation of reference values into suitable groups, analysis of the spreading of individual group, identification of outliers, and determination of reference limit.

7.3.1   *PARTITIONING OF REFERENCE VALUES*

It is termed as stratification, categorization, or subgrouping also. Its results are called as partition, strata, category, class, or subgroup. For example, the subset of reference data points and the corresponding reference values might be partitioned on the basis of age sex, and other characteristics. The aim of partitioning is to reduce, if possible and necessary, variation among subjects to minimize biological "noise."

7.3.2   *INSPECTION OF DISTRIBUTION*

It is suggested to display the reference distribution graphically, and its subsequent inspection. A histogram should be prepared manually or by a computer program. The examination of histogram is a safeguard against the misapplication or misinterpretation of statistical methods; it might provide valuable information about data.

7.3.3   IDENTIFICATION AND HANDLING OF OUTLIERS

An outlier is an erroneous value that deviates significantly from proper reference values. The identification of random outliers by visual evaluation of a histogram is a reliable method. However, the inspector must follow that values near the furthest point on the long tail of a skewed distribution easily may be misinterpreted as outliers. If the distribution is positively skewed, the evaluation of a histogram displaying the logarithms of values might aid in the identification of outliers.

7.3.4   DETERMINATION OF REFERENCE LIMITS

In clinical routine, an observed patient's value is usually compared with its equivalent reference interval and is defined by a couple of reference limits. The interval may be understood in different ways and is a useful condensation of information carried by the total set of reference values. The reference limit describes its distribution and provides information on the expected variation of values in the selected set of reference individuals. Diversely, the clinical decision limit provides information on optimal separation among clinical categories. Categories of reference intervals include tolerance, prediction, and interpercentile interval.

The interpercentile interval can be concluded on the basis of parametric and nonparametric statistical techniques. The parametric method for the identification of percentile and their confidence intervals assumes a certain type of distribution. It is based on estimates of population parameters such as mean and standard deviation (SD). The majority of parametric methods are based on Gaussian distribution. For example, a parametric method is used if the true distribution is believed to be Gaussian, and reference limits (percentiles) are determined as the values located two SDs below and above the mean. Several nonparametric methods are available, but those based on ranked data are simple and reliable. These methods allow nonparametric estimation of confidence interval of the percentile. The nonparametric method does not conclude assumptions pertaining to the type of distribution and do not utilize estimates of distribution parameters. The percentiles are determined simply by removing the required percentage of values in each edge of the subset reference distribution. On comparison of the results obtained by these two methods, the estimates of the percentiles usually are very similar (Burtis and Bruns, 2014). Simple and reliable nonparametric method, especially in its bootstrap version, generally is preferable to the parametric method. The bootstrap method is an extension of nonparametric method. However, this principle may be employed with both parametric or nonparametric.

7.4   USE OF REFERENCE VALUES

Interpretation of medical laboratory data requires a comparison of the patient's values with the reference values. An observed value (patient's value) may be compared with reference values. The comparison is often similar to hypothesis testing, but in strict sense, it is a seldom statistical testing. It is suggested to consider the reference values as the yardstick for a less formal assessment than hypothesis testing. A convenient presentation of the observed value and the reference interval on the same report sheet may be helpful for the busy clinician or healthcare provider. For example, the reference intervals may be preprinted on report forms, or the computer system may select the appropriate age and sex-specific reference interval from the database and print it next to the test result or in graphical form. An observed value might be categorized as low, usual, or high (three classes), depending on its position in relation to the reference interval. On reports, a convenient practice is to flag unusual results (e.g., through use of the letters L and H for low and high, respectively). Another popular method of classification is to express the observed value by a mathematical distance measure. For example, the well-known SD-unit, or normal equivalent deviation. It is calculated as the difference between observed value and the mean of the reference values divided by their SD. This measure, however, is unreliable if the distribution of values is skewed (Burtis and Bruns, 2014).

7.5   INTERPRETATION OF EXPERIMENTAL RESULTS

A clinical laboratory should produce a precise and dependable result. Different facts are involved in the accurate interpretation of test results (Figure 7.1). The smallest amount of sample which could be perceived by a particular clinical test is referred as detection limit and it defines the *analytical sensitivity* of the test. If a particular disease marker is elevated in both severe and early stages of the disease, the method for the detection of marker will useless if it does not detect low levels, as an early diagnosis would not be possible. Test's *Analytical specificity* refers to whether any other similar moiety interferes with the test or not. For example, in diabetic diagnosis, it is essential that the test should be able to differentiate between insulin and proinsulin; otherwise, the test will give plausible elevated results. Thus, the clinical test should be analytically sensitive and specific. *Precision, accuracy,* and *bias* are the basic terms used to analyze certain parameters of clinical tests (Figure 7.1). Precision tells how close the repetitive points of the same sample lie, accuracy explains how close the value expected is to the actual value and bias describes the variables which deviate the precision and accuracy leading to over and under-reporting or large random background changes (Basten et al., 2010) (Figure 7.1).

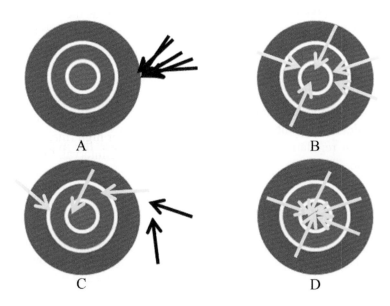

FIGURE 7.1 (A) detailed but not accurate, (B) accurate but not detailed, (C) Neither precise nor accurate, and (D) detailed and accurate clinical test results.

7.6 QUALITY MANAGEMENT

Quality is defined as conformity to the requirements of user or customer satisfaction and expectations. The general principles of TQM are customer focus, management commitment, training, process capability and control, and measurement using advanced tools for the enhancement of quality enhancing. Accurate and on-time test reports are the accountability of the laboratory. However, numerous difficulties arise in advance and after the submission of the specimen (Burtis and Bruns, 2014). Therefore, appropriate management of the entire testing process has to be done before analysis during and after the analysis. Parameters before analyzing the process are discussed as:

- Test usage and practice guidelines;
- Patient identification;
- Turnaround time (TAT);
- Laboratory logs;
- Transcription errors;
- Patient preparation;
- Specimen collection, transport, division, and spreading of aliquots.

7.6.1   TEST USAGE AND PRACTICE STRATEGIES

Traditionally, laboratory test usage always has been monitored or controlled to some degree, but the current emphasis on the cost of medical care and government regulation of medical care may increase the importance of this factor.

7.6.2   PATIENT IDENTIFICATION

Accurate identification of patient specimen is a prominent concern for laboratories. Major, an increase in error rate occurs when handwritten labels and request forms are used. The use of barcoding technology for patient identification can minimize this error to a greater extent.

7.6.3   TURNAROUND TIME (TAT)

The elapsed time since the beginning of a test and till the test result has been reported is termed as TAT of a given test. Delayed and lost requisition forms, samples, and unlabeled reports are unacceptable TATs. So, in practice, it is essential to record the real-time during collection, analysis, and conclusion of test results.

7.6.4 LABORATORY LOGS

When a blood sample arrives at laboratory, a request/report form usually should be accompanied by them. Information such as name of the patient, identification number and the tests demanded on the form needs to be checked on the label of the specimen tube should be confirmed. Also, the specimen should be inspected for adequacy of volume and constraints that hinder the assay procedure, such as lipemia or hemolysis. The specimen should be stored properly, and the identification of data and arrival period should be recorded in the main log.

7.6.5 TRANSCRIPTION ERRORS

In laboratories, lack of electronic identification and tracking system, a considerable threat of transcription error in manual data entry, even the results are checked twice. Computerization reduces this transcription error because computerized systems have error-detection routines programmed into the terminal entry function. These routines may include digits, limit, test-correlation, and verification checks with the master hospital files. There are many physical factors that affect the laboratory test related to the patient such as immediate intake of alcohol, drugs, or food and smoking, exercise, stress, sleep, posture at the time of specimen collection, and various other factors. Preparing patients for a particular test is essential to obtain rightful results. The laboratory must follow the instruction and procedure for the proper preparation of patients and as well as for specimen acquisition.

7.6.6 SPECIMEN COLLECTION

The technique used to acquire a sample may also affect laboratory tests. Use of unsuitable vessels and improper preservatives can affect the test results. One of the solutions to counter this problem of laboratory processing is to assign specially trained laboratory personnel for the specimen collection.

7.6.7 SPECIMEN TRANSPORT

Proper transportation of samples to local, regional, and high-end laboratories has a considerable effect on the stability of samples. To avoid the errors during the transportation, the concerned labs should reject specimens that arrive in an unsatisfactory condition (such as a thawed specimen that should have remained frozen).

7.6.8 SPECIMEN SEPARATION AND DISTRIBUTION OF ALIQUOTS

Separation of blood samples and aliquot circulation are the functions performed during analysis and are under the regulation of laboratory. The key factors that govern the process are centrifuges, containers, and personnel. Centrifuges should be monitored for speed, time, and temperature. Collection tubes, pipettes, stoppers, and aliquot tubes are source of calcium and trace metal contamination; each lot number of material used should be tested for contamination of calcium and other possible elements.

KEYWORDS

- aliquots
- laboratory logs
- outliers
- standard deviation
- transcription errors
- turnaround time

REFERENCES

Basten, T., Van, B. E., Geilen, M., Hendriks, M., Houben, F., Igna, G., Reckers, F., De Smet, S., Somers, L., Teeselink, E., & Trčka, N., (2010). Model-driven design-space exploration for embedded systems: The octopus toolset. In: *International Symposium on Leveraging Applications of Formal Methods, Verification and Validation* (pp. 90–105). Springer, Berlin, Heidelberg.

Burtis, C. A., & Bruns, D. E., (2014). *Tietz Fundamentals of Clinical Chemistry and Molecular Diagnostics-E-Book.* Elsevier Health Sciences.

CHAPTER 8

Cell Membrane Receptor

PREM PRAKASH KUSHWAHA, SWASTIKA DASH,
ATUL KUMAR SINGH, and SHASHANK KUMAR

*School of Basic and Applied Sciences, Department of Biochemistry,
Central University of Punjab, Bathinda, Punjab–151001, India,
Tel.: +91 9335647413,
E-mail: shashankbiochemau@gmail.com (S. Kumar)*

ABSTRACT

Cellular membrane receptors occur onto the cell plasma membrane and sometimes on the intracellular membrane. These cellular receptors receive signals and facilitate various biological responses. Receptors such as ion channel linked receptors (ICLRs), enzyme coupled receptors (ECRs), tyrosine/serine/threonine kinase receptors, guanylyl cyclase receptors and globular protein-coupled receptors mainly regulates all the signal transduction processes. Interaction related studies of the signal and receptor can be achieved by the numerous assays like cell behavior assays, pull-down assay, activity assay, signal transduction event visualization, qualitative alteration in calcium and pH, immuno-histochemistry, FRET, FRAP, western blot, immunoprecipitation, and constructed fusion proteins.

8.1 INTRODUCTION

Cell membrane receptors and quantitative ligand binding "communication and cooperation" are must for all multicellular organisms to sustain homeostasis at the cellular level. Cell signaling administrates basic activities between two different cells or cells and its external environment and coordinates their crucial activity. Widespread mechanism of cellular control involves cell-to-cell communication. Extracellular signaling molecules regulate the metabolic processes, tissue growth, and their differentiation, protein trafficking and also intracellular and extracellular fluid maintenance within an organism. A eukaryotic system like yeast and protozoan secretes a molecule termed pheromones. Pheromones mediate cell aggregation for sexual mating or differentiation. In most of the signaling pathways, ligand interaction or binding to receptor results in activation of transcription factor in the cytosol, allowing them to move into the nucleus and fuel (or infrequently repress) specific target genes to undergo transcription. Otherwise, receptor activation stimulates cytosolic protein kinases which translocate inside the nucleus and control nuclear transcription factors. Cell signaling generally involves chemical messengers such as protein and steroids. Signaling may be either local or at long distance. Cell recognition proteins such as glycolipids and glycoproteins (e.g., blood type proteins) are attached to cell exterior. Local regulators may be paracrine (secreted signal like growth factors) or synaptic (directed signal like neurotransmitter) in nature. Long distance signaling mostly involve hormone and occurs indistinctive system of the organism such as:

- Glands releases animal hormones into the bloodstream. The hormones travel distance in body and cause changes in a bunch of cells at almost the identical potential (for example adrenaline).
- In animal, nervous system involves long-distance communication through electrical signals sent via neuron.
- Endocrine System in animals also perform long distance communication. Lymph nodes, pituitary gland, adrenal gland etc. produce hormones inside the cells or pour it into the bloodstream. In the plant system hormone travel through vascular system, plasmodesmata, or sometimes released into the air (e.g., ripening fruit).

The interaction among the two given cells in the body may be explained in the term of cell junctions. Cell junctions may consider as a tunnel of protein that directly links the adjacent cells. They allow material (chemical, water, etc.) to pass through one cell to another. Cells also have guard against unfamiliar cells and invaders. It provides safety from the unnecessary signals. They may be membrane-associated cell surface molecules such as glycoproteins or glycolipids. Prokaryotic and eukaryotic systems transfer information by various physical and chemical stimuli. Chemical signals generally originated in the animal cell may be of three types: autocrine, paracrine, and endocrine.

8.1.1 AUTOCRINE

In autocrine, signaling, animal cell secretes chemical messenger or hormone that binds to their perspective autocrine receptor on the similar cell, resulting in cellular response.

8.1.2 PARACRINE

In paracrine signaling, a cell produces a signal, to convince changes in the neighboring cells resulting in a change in their response.

8.1.3 ENDOCRINE

Endocrine signaling transfers their signals (such as secretion of hormones inside the circulatory system) to regulate distant tissues.

In all the three circumstances, signals are generally identified by a specific receptor. Cellular responses may include change in gene expression, activation of DNA synthesis, apoptosis, cytoskeleton rearrangement, metabolic enzyme regulation, etc.

8.2 RECEPTORS FOR CELL SIGNALING

According to the characteristic signal, target cell responses by a protein called a receptor. Receptors are the sensing element in the system of signal connections that coordinate the purpose of all the diverse cells in the body (Figure 8.1). Receptor binds to ligand with precise specificity and initiates a reaction in the target cell. In the eukaryotic system, the receptor may be confined in the plasma membrane or cell cytoplasm. Plasma membrane receptor generally receives polar signals. Plasma membrane has three types of receptor as follows:

- Ion channel linked receptors (ICLRs);

- Enzyme coupled receptors (ECRs); and
- Globular protein-coupled receptors (GPCRs).

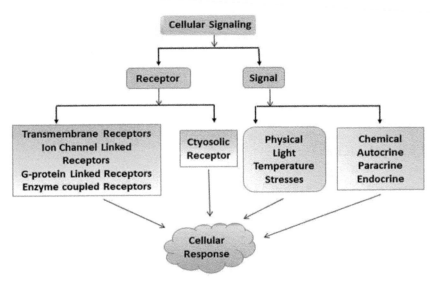

FIGURE 8.1   Cell signaling receptors and signals involved in signal transmission.

8.2.1 *ION CHANNEL LINKED RECEPTORS (ICLRS)*

Ion channel-linked receptor interacts with a ligand and changes its structural conformation to expose pore in the membrane that permits the movement of ion across the membrane. For the requirement of channel formation, ICLRs possess a wide-ranging membrane-spanning region. To cooperate with the phospholipid fatty acid tails (form the core of the plasma membrane) the majority of amino acids lying in the membrane-spanning region tend to be hydrophobic in nature. On the contrary, the amino acids that lie in the inner side of the channel are hydrophilic in nature and they allow the movement of ions and water across the membrane. When a ligand interacts with the channel's extracellular part, a conformational transformation in the protein arrangement takes place. It permits ions like calcium, sodium, magnesium, and hydrogen to surpass through. Every kind of ion channel receptor has its own signal. Acetylcholine receptor positioned in the vertebrate skeletal muscle cell plasma membrane is an example of ICLR.

8.2.2   ENZYME COUPLED RECEPTORS (ECRS)

Enzyme-linked receptors reside on the cell-surface with intracellular domain allied with an enzyme. Sometimes, the receptor's intracellular domain itself plays as an enzyme or the ECR has an intracellular domain that binds openly with an enzyme. The ECR contains large intracellular and extracellular domains of the membrane-spanning portion contain a single alpha-helical part of the peptide strand. When ligand binds to the extracellular domain it results in signal transmission via the membrane and triggers the enzyme, which starts a string of proceedings inside the cell that ultimately results in a cellular response. ECRs are divided into six major groups:

8.2.2.1   RECEPTOR TYROSINE KINASES

This type of receptor has kinase activity, used in the phosphorylation of some tyrosine residues specifically, present on their own surface. They sometime phosphorylate target proteins also.

8.2.2.2   TYROSINE KINASES ASSOCIATED RECEPTORS

These receptors directly activate cytoplasmic tyrosine kinases. They do not have the intrinsic kinase activity. This type of receptors is enormously infrequent in plants. Numerous tyrosine kinases associated receptors are epidermal growth factor (EGF), platelet-derived growth factor receptor (PDGFR), insulin-like growth factor receptor (IGFR), fibroblast growth factor receptor (FGFR), vascular endothelial cell growth factor receptor (VEGFR), Ephrin receptor (ER), nerve growth factor receptor (NGFR), etc.

8.2.2.3   RECEPTOR SERINE/THREONINE KINASES

Receptor serine/threonine kinases phosphorylate specific serine/threonine on themselves and various gene regulatory proteins. These kinases are the leading group of receptor families present in plants. Antigen receptors, interleukins on lymphocytes, integrin's receptor, cytokine receptors, growth/prolactin receptors, etc. belong to receptor serine/threonine kinase family.

8.2.2.4 HISTIDINE KINASES ASSOCIATED RECEPTORS

Histidine kinases associated receptor activates two-component signaling system pathways.

8.2.2.5 RECEPTOR GUANYLYL CYCLASE

In general, there are three guanylyl cyclase cell receptors. Commonly these receptors emerged to impart in the regulation of fluid movement or volume. Nitric oxide concentration is basically regulated by the heterodimeric form of guanylyl cyclase and also seems to be regulated by $Ca^{2+}/$ calmodulin-sensitive nitric oxide synthase.

8.2.2.6 RECEPTOR-LIKE TYROSINE PHOSPHATASES

This kind of phosphatases removes the phosphate group from tyrosine in some specific target proteins.

8.2.3 GLOBULAR PROTEIN-COUPLED RECEPTORS (GPCRS)

GPCRs superfamily includes the major and most varied assembly of proteins in mammals. Robert Lefkowitz and Brian Kobilka got the Nobel Prize in Chemistry in 2012, for the groundbreaking discovery that discloses the internal working of a significant family of G-Protein Coupled Receptor family

GPCRs are commonly called serpentine receptors, seven-transmembrane domain receptors, heptahelical receptors, 7TM receptors, and G protein-linked receptors (GPLR). There are >800 GPCRs encoded by the human genome. GPCRs mainly involved in the transfer of information from outside to inside of the cell. GPCRs are accountable for every facet of human biology such as visual sense, gustatory sense (taste), sense of smell, behavioral, and mood regulation, regulation of immune system activity and inflammation, autonomic nervous system transmission, cell density sensing, homeostasis modulation and also involved in growth and metastasis of some types of tumor. In the current scenario, drug discovery efforts endeavor at both improving therapies for greater than 50 established

GPCR targets, and at escalating the list of targeted GPCRs (Lappano and Maggiolini, 2011) (Figure 8.2).

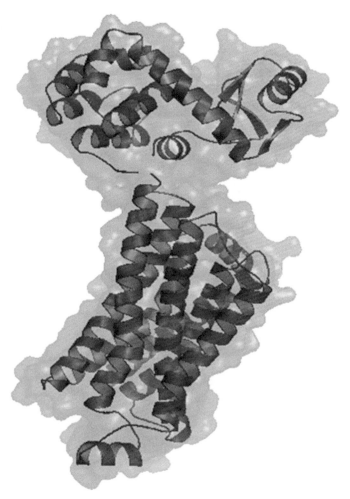

FIGURE 8.2   Human B2-adrenergic G protein-coupled receptor (PDB: 2RH1) interacts with epinephrine and mediates physiologic response like smooth muscle relaxation.

Up to date, ~45% of all pharmaceutical drugs are known to target GPCRs. There are several kinds of receptors associated with GPCRs. Some common examples are $GABA_B$ receptors ($GABA_BR_1$ and $GABA_BR_2$), taste receptors (T_1R_3 and T_1R_2), adrenergic receptors (three subfamilies α_1,

α_2 and β) (Figure 8.2), opioid receptors (three cloned subtypes δ, κ, and μ), somatostatin receptors (five subtypes $SSTR_{1-5}$), purinergic receptors (neurotransmitters in the CNS, immune system, CVS, and some different tissues i.e., ATP, and adenosine) and olfactory receptors (largest GPCRs family, about >300 members oxytocin, vasopressin, and other receptors). Based on the similarity of sequence within the segments of 7TM, they are categorized into five families (Fredriksson et al., 2003).

- Rhodopsin family with 701 members;
- Adhesion family with 24 members;
- Frizzled/taste family with 24 members;
- Secretin family with 15 members; and
- Glutamate family with 15 members

Most commonly, GPCRs activating ligands are light-sensitive compounds, odors, pheromones, hormones, and neurotransmitters, and differ in size from tiny molecules to peptides and large proteins. The ligand binds in the inner nook between helices leading their minor angular dislocation, visible as a conformation alteration in the cytoplasmic domain. Cytoplasmic section of the receptor binds a three-component protein complex associated with that comprise tightly bound GDP (heterotrimeric G- protein). Hence, also known as G-protein coupled receptor (GPCR) family. The largest subunit 'α' contains the bound GDP and is bilayer lipid-anchored. There are several variants of 'α' such as $G\alpha_s$, $G\alpha_i$, and $G\alpha_q$, etc. The heterotrimer is accomplished by $G_{\beta\gamma}$ subunits, which binds to each other very tightly. The small gamma subunit is anchored in lipid by a farnesyl chain. In the absence of ligand, bound GDP cannot be released from G protein, resulting in inactivation of the protein.

8.3   QUANTITATIVE FEATURES OF BINDING OF RECEPTOR AND LIGAND

8.3.1   LIGAND-BINDING AND CELL-SIGNALING STUDIES

Different assays have been established to study the kinase activity and signaling pathways. Dependent on the phase of the study several known techniques have the potential to explore valuable information regarding cell-signaling. In the following divisions, we are going to discuss some precise approaches used for the study of cell signaling.

8.3.2 INHIBITOR BLOCKS SPECIFIC KINASES

Several inhibitors are currently acknowledged to study the specific interactions with receptor or specific kinases. These inhibitors may be exact for single kinase at low doses. High doses of these inhibitors may distress wider class of molecules. Most of the inhibitors act as competitive inhibitor of ATP binding site and are reversible. To avoid the distress of wider class of molecules, persistent levels of inhibitor is compulsory throughout the experimental duration. Evaluation of inhibitor potential should consist of biochemical, morphological, and behavioral analysis.

8.3.3 CELL BEHAVIOR ASSAYS

Cell behavior assays may define alterations in characteristics such as changes in cell shape, cytoskeletal elements, matrix binding, migration, or differentiation. Spreading, attachment, and migration assays are commonly being used to regulate up or down signal transduction proteins and their respective change in the cellular behavior (Berrier et al., 2000). Inhibitors also regulate signal transduction proteins by precise inhibitors of cell surface receptors and intracellular kinases. Transfecting or microinjecting the cells with plasmids can also explore changes in the cellular behavior by altering the activity of specific proteins/enzymes (Berrier et al., 2000). Blocking protein synthesis by antisense oligonucleotides specific to the mRNA of that specific protein can modulate function of the specific protein. Studies based on this technology showed that different signal transduction pathway contribute to cytoskeletal rearrangements. The ECM-stimulated fluctuations in the arrangement of actin cytoskeleton have been well recognized in tissue by using transmission electron microscopy (TEM) and confocal microscopy (Chu et al., 2000). Researchers also perform *in vitro* wound-healing method in which the cells grown to confluence and then a scrape/wound is positioned in the culture dish (Song et al., 2000). The cells are detected by stirring the surface of the wound under numerous conditions to govern what proteins are essential for cell migration resulting into wound healing.

8.3.4   SIGNAL TRANSDUCTION EVENT VISUALIZATION

Usually, signal transduction activities like pH or change in calcium levels changes can be detected through intracellular fluorescent indicators. Techniques like fluorescent markers, immunohistochemistry, and caged proteins can be utilized to study translocation of specific protein in a definite plasma membrane lipid bilayer. Recently, transportation of proteins has been tracked by integrating a fluorescent protein gene (green fluorescent protein (GFP)) into genetic vector encrypting the protein to be studied.

8.3.5   QUANTITATIVE ALTERATION IN CALCIUM AND pH

Alteration in calcium or pH levels can be accomplished by single-wavelength dyes. These fluorescent probes show a spectral reply upon binding with Ca^{2+}. The alteration in concentrations of intracellular free Ca^{2+} can be detected by using FACS (fluorescence activated cell sorting) and fluorescence spectroscopy. Mostly fluorescent indicators are derivatives of Ca^{2+} chelators (ethylene glycol-bis (β- aminoethyl ether)-N,N,N,'N'-tetraacetic acid), o-aminophenol-N,N,O-triacetic acid, and 1,2-bis(o- aminophenoxy) ethane-N,N,N',N'-tetraacetic acid. The mutual ratiometric dye for Ca^{2+} is indo-1 and fura-2. This dye alters the wavelength within a range of ion concentration. Single wavelength dyes are commercially available such as calcium orange, fluo-3, calcium green, fura red, calcium crimson, and Rhod-2 (Haugland and Johnson, 1999). These compound needs a fluorescent microscope with the capability to record moderately fast alteration in emission wavelength by using charged coupled device (CCD) cameras, enhanced video or very fast confocal microscopes.

8.3.6   IMMUNOHISTOCHEMISTRY

To locate the signaling proteins, researchers are using immunohistochemistry technique. This technique indicates not only the protein activation but also might confirm the state of the target protein. Many companies are also manufacturing antibodies which signal the activated state of protein. These antibodies have recognized epitope in phosphate or other activated conformation. Anti- active (antibodies against active epitopes) antibodies are also accessible for the specific signal transduction protein. In rare cases when the anti-active antibody is unavailable, a substitute method is used. In this method, the cells are twice labeled with a target antibody and another antibody that identifies all the phosphorylated amino acids like serine, threonine, or tyrosine. The outcome of method is an overlap of both the signals i.e., a site where the active protein occurs and the site where the single-labeled protein is not activated.

8.3.7 CONSTRUCTED FUSION PROTEINS

Living cells can also be identified with GFP-tagged proteins. Vectors with GFP attached to the protein of interest are transfected into tissues, cells, or transgenic animals to follow the pattern of protein expression. GFPs has been altered to produce several emission wavelength like yellow FP (YFP), cyan FP (CFP) and DsRed so that it could be utilized in combination, to label the multiple proteins (Ayoob et al., 2001). GFP-tagged fusion protein transfection is a predominant method, if the GFP tagged protein alters cellular location after activation. In a study, a section of MBP (myelin basic protein) that possesses a single consensus PRTP (ERK/MAP kinase phosphorylation motif) was combined with the GFP. The protein that has been fused and transfected inside the mammalian cell behaves as a substrate kinase. The GFP-MBP fused protein gets phosphorylated following serum stimulation while a MEK inhibitor obstructed this deviation in phosphorylation (Mandell and Gocan, 2001).

8.3.8 FRET AND FRAP

FRAP (fluorescence recovery/redistribution after photobleaching) and FRET (fluorescence resonance energy transfer) require protein/organelle to be studied with fluorescently labeled either in live cells or in fixed preparations. FRET is the nonradioactive shift in energy from a donor that is in an excited state to an adjacent acceptor (Mátyus, 1992). A fluorescent molecule that gets excited by a particular wavelength of light is the donor here. The donor emits a higher light wavelength that excites the fluorescence acceptor molecule. Most of the fluorescent molecules can be used as a donor/acceptor pair. Donor/acceptor pair examples are the Cy3, Cy5, CFP, and YFP. Energy transfer is reliant on the space between the fluorescent molecules. As the acceptor sinks in the donor fluorescence, the donor will quench the energy and its lifetime will decrease. FRET is a very influential light microscopic technique to determine whether two proteins are within 10–70Å of each other rather than co-localized with confocal microscopy.

FRAP determines the energy kinetics of diffusion through cells or tissues. It is accomplished by measuring the lateral diffusion in two dimensions, of a thin film consisting of probes that are fluorescently labeled or to scan single cells. Before the targeted region is photo-bleached, the cells are formerly loaded with the fluorescent molecule of concern and the movement of fluorescent molecules back into the bleached part is measured (Mochizuki et al., 2001). This system permits the determination of the diffusion and mobility of small molecules in the living cells' cytoplasm. It also records the movement of macromolecules such as RNA, heavy protein or drug outside and inside the cell organelles (nucleus).

8.3.9   WESTERN BLOTS

In immune-genetics and molecular biology, the protein immunoblot or western blot is a broadly used technique to analyze specific proteins from an extract or tissue homogenate. The benefit of this kind of blotting is that it specifies all of the proteins that may be tyrosine phosphorylated and can document how to separate proteins that may increase or decrease the signal. The blot does not identify the precise protein. Either a sister blot can be probed with the definite antibody or the same blot can be exposed and re-probed with additional primary antibody for the identification of the single protein.

8.3.10   IMMUNOPRECIPITATION

Immunoprecipitation is comparatively easy and involves the same apparatus as western blot. Beads are made-up of a diversity of substances. Studies showed that cross-link primary antibody of a specific protein or all tyrosine-phosphorylated proteins to the beads may occur. The cells are lysed in a buffer comprising protease inhibitor and incubated with the antibody-coated beads. The proteins get separated by using SDS electrophoresis and recognized by the same procedures of western blot. The protein-protein interactions confirmed by western blots and immune-precipitation from cell lysate may not disclose the actual situation *in vivo* thus additional experiments are essential to determine precise interactions.

8.3.11 GST BINDING OR "PULL-DOWN" ASSAYS

The glutathione S-transferase (GST) binding or "pull-down" assay is equivalent to immunoprecipitation. It also detects direct protein-protein interactions. Trash of proteins is produced through a bacterial expression system with a GST tag. In a study, the researcher has used GST-labeled proteins like Rho GDP, Rho GTP, and the domain which binds Rho in protection (RBD-GST), and proper GST control verified that the corneal epithelial cells show a biphasic reaction to collagen. A reduction in Rho GTP after 15 min of collagen stimulation was done and after that, an augmentation at 30 min similar to the response in endothelial cells was carried out (Ren et al., 2000).

8.3.12 ACTIVITY ASSAYS

To identify the activity of a particular kinase, the activity assay exposes the cell lysate to a known substrate for the enzyme in the existence of radioactive phosphate. The partition of the product is accomplished by electrophoresis and bared to the x-ray film to detect the incorporated protein to the isotope. Apoptosis results in susceptible cells to undertake a cascade of morphologic and enzymatic changes. Various signal transduction pathways, as well as several specific degradative enzymes, become activated during apoptosis. These enzymes may chop crucial structural apparatus of the cell containing small nucleoproteins, actin cytoskeletal elements as well as nuclear lamins (De Laurenzi and Melino, 2000). To establish that the cells encompass active caspase-3 or not, caspase-3 substrates consist of caspase-3 recognition sequence DEVD (aspartic acid-glutamic acid valine-aspartic acid) was used in a fluorophore-derivatized peptide which behaves same as the structural loop conformation existing in the globular proteins' indigenous protease cleavage sites. The compound is not fluorescent unless it is chopped by endogenous caspase-3. To check the reaction is particular for the caspase-3 activity. Certain tissues can be formerly treated with the caspase-3 inhibitor such as Z-VAD-FMK.

KEYWORDS

* **enzyme coupled receptors**
* **Ephrin receptor**
* **epidermal growth factor**
* **fibroblast growth factor receptor**
* **fluorescence-activated cell sorting**
* **heterotrimeric G-protein**

REFERENCES

Ayoob, J. C., Shaner, N. C., Sanger, J. W., & Sanger, J. M., (2001). Expression of green or red fluorescent protein (GFP or DsRed) linked proteins in nonmuscle and muscle cells. *Mol. Biotechnol.*, *17*(1), 65–71.

Berrier, A. L., Mastrangelo, A. M., Downward, J., Ginsberg, M., & LaFlamme, S. E., (2000). Activated R-Ras, Rac1, PI 3-kinase and PKCε can each restore cell spreading inhibited by isolated integrin β1 cytoplasmic domains. *J. Cell Biol.*, *151*(7), 1549–1560.

Chu, C. L., Reenstra, W. R., Orlow, D. L., & Svoboda, K. K. H., (2000). ERK and PI-3 kinase are necessary for collagen binding and actin reorganization in corneal epithelia. *Invest. Ophthalmol. Vis. Sci.*, *41*(11), 3374–3382.

De Laurenzi, V., & Melino, G., (2000). Apoptosis: The little devil of death. *Nature*, *406* (6792), 135, 136.

Fredriksson, R., Lagerström, M. C., Lundin, L. G., & Schiöth, H. B., (2003). The G-protein-coupled receptors in the human genome form five main families. Phylogenetic analysis, paralogon groups, and fingerprints. *Mol. Pharmacol.*, *63*(6), 1256–1272.

Haugland, R. P., & Johnson, I. D., (1999). Intracellular ion indicators. *Fluorescent and Luminescent Probes for Biological Activity* (pp. 40–50). Mason WT. Academic Press, Cambridge.

Lappano, R., & Maggiolini, M., (2011). G protein-coupled receptors: Novel targets for drug discovery in cancer. *Nat. Rev. Drug Discov.*, *10*(1), 47–60.

Mandell, J. W., & Gocan, N. C., (2001). A green fluorescent protein kinase substrate allowing detection and localization of intracellular ERK/MAP kinase activity. *Anal. Biochem.*, *293*(2), 264–268.

Mátyus, L., (1992). New trends in photobiology: Fluorescence resonance energy transfer measurements on cell surfaces. A spectroscopic tool for determining protein interactions. *J. Photochem. Photobiol. B.*, *12*(4), 323–337.

Mochizuki, N., Yamashita, S., Kurokawa, K., Ohba, Y., Nagai, T., Miyawaki, A., & Matsuda, M., (2001). Spatio-temporal images of growth-factor-induced activation of Ras and Rap1. *Nature*, *411*(6841), 1065–1068.

Ren, X. D., Kiosses, W. B., Sieg, D. J., Otey, C. A., Schlaepfer, D. D., & Schwartz, M. A., (2000). Focal adhesion kinase suppresses Rho activity to promote focal adhesion turnover. *J. Cell Sci.*, *113*(20), 3673–3678.

Song, Q. H., Singh, R. P., Richardson, T. P., Nugent, M. A., Trinkaus-Randall, V., (2000). Transforming growth factor-β1 expression in cultured corneal fibroblasts in response to injury. *J. Cell Biochem.*, *77*(2), 186–199.

CHAPTER 9

Mechanisms of Signal Transduction

PREM PRAKASH KUSHWAHA,[1] REBATI MALIK,[1]
P. SESHU VARDHAN,[2] SHIV GOVIND RAWAT,[3] SHARMISTHA SINGH,[4]
and SHASHANK KUMAR[1]

[1]*School of Basic and Applied Sciences, Department of Biochemistry, Central University of Punjab, Bathinda, Punjab–151001, India, Tel.: +91 9335647413, E-mail: shashankbiochemau@gmail.com (S. Kumar)*

[2]*School of Biotechnology, Jawaharlal Nehru Technological University, Kakinada–500085, Telangana, India*

[3]*Department of Zoology, Banaras Hindu University, Varanasi–221005, Uttar Pradesh, India*

[4]*Department of Biochemistry, University of Allahabad, Allahabad, India*

ABSTRACT

Signal transduction controls all the homeostasis and cellular function inside the body. Signaling pathways start with the interaction of the signal molecule with the receptor onto the membrane or directly to the target inside the cytosol. Some signaling pathways also start with autophosphorylation of the receptor. Signal molecule transfers its information in a sequential manner and regulates target gene expression inside the nucleus. Expression of targeted genes directs cellular functions such as motility, differentiation, cell proliferation, and survival. In this chapter, we emphasize the different signaling such as GPCRs, Wnt, notch, sonic hedgehog (SHH), JAK-STAT, NF-kB, and TGF-β signaling.

9.1   INTRODUCTION

The three primary cell signaling stages are reception, transduction, and response (Figure 9.1). Reception receives the signals that come from the other cells or extracellular environment. The signaling molecules referred as ligand bind to the specific receptor protein. Transduction provides signal mediated flow of information inside the cell. Secondary messenger transduces the original exterior signal inside the cell. Based on cell signal transduction cell signaling is also known as signal transduction pathway. Lastly, response stands for cell response after receiving the signals, resulting into protein synthesis, energy production, or the cell enters into mitosis.

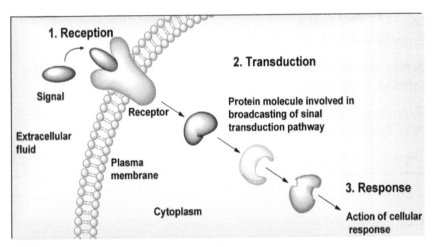

FIGURE 9.1   An overview of cellular signaling.

In this chapter, we will focus on the following major signaling pathways (Table 9.1):

- GPCRs (globular protein-coupled receptors);
- Wnt signaling;
- Notch signaling pathway;
- Sonic hedgehog (SHH) pathway;
- JAK-STAT pathway;
- NF-kB pathway;
- TGF-β signaling.

TABLE 9.1 A Look of the Major Receptor Classes and Signaling Pathways in Mammalian Systems

Receptor Class/ Pathway	Ligand	Receptor	Signal Transduction
GPCRs	Epinephrine Glucagon Serotonin Vasopressin ACTH Adenosine	Seven transmembrane α helices, cytosolic domain combined through a membrane-bound trimeric G protein	Second-messenger pathways implicates cAMP or IP3/DAG, Ion linked channels, MAP kinase pathway
Wnt	Secreted Wnt	Frizzled (Fz) has seven transmembrane α helices; associated membrane-bound LDL receptor-associated protein (Lrp) mandatory for receptor activity	Assembly of the multi-protein complex at membrane which inhibits proteasomal degradation of cytosolic β-catenin transcription factor
Notch/Delta	Membrane- bound Delta (δ) or serrate protein	Extracellular subunit of Notch receptor non-covalently associated with transmembrane- cytosolic subunit	Intramembranous proteolytic cleavage of receptor's transmembrane domain with the discharge of cytosolic portion that performs role of co-activator for nuclear transcription factors
Hedgehog	Cell-tethered hedgehog	HH associates to Patched (Ptc) protein comprised of 12 transmembrane α helices; activation of signaling from smoothened (7 transmembrane α helices) activates the signaling	Releasing of a transcriptional stimulator by proteolytic cleavage from the multi- protein complex in cytosol.
Cytokine receptors	Interferons Erythropoietin, Growth hormone IL-2, IL-4 Other cytokines	A single transmembrane α helix. Extracellular domain contains a conserved multi-strandfold, JAK kinase associate with the intracellular domain	Direct activation of STAT transcription factors in the cytosol, PI-3 kinase pathway, IP3/DAG pathway, Ras-MAP kinase pathway
NF-kB	TNF-α IL-1	Various in mammals	Inhibitor protein gets degraded by the phosphorylation with the release of active NF-B transcription factor in the cytosol
TGF	TGF BMPs Activin Inhibins	Intrinsic protein serine/threonine kinase action possessed by cytosolic domain, i.e., Type I and II receptors.	Direct stimulation of cytosolic Smad transcription factors C

9.2 GLOBULAR PROTEIN-COUPLED RECEPTORS (GPCRS)

There are two main signal transduction pathways that comprise GPCR (G-protein-coupled receptors)

- cAMP pathway; and
- Phosphatidylinositol (PI) pathway.

9.2.1 CYCLIC AMP PATHWAY

Ligand-bound receptor serves as guanosine exchange factor (GEF) and permits GTP. There is a conformational change when Gαs binds GTP and its affinity for the receptor and G_β is lost. So the association between whole complexes breaks up and Gαs separates. Because of its lipid anchor ability Gαs-GTP, remain attached to the membrane. The affinity of GTP increase for another membrane- associated protein i.e., adenylate cyclase and binding of Gαs-GTP excites adenylate cyclase to produce cyclic AMP (Figure 9.2). A twelve transmembrane glycoprotein adenylyl cyclase catalyzes the synthesis of cAMP from ATP in the presence of Mg^{2+}or Mn^{2+} cofactors. Gαs represents the stimulatory G protein. Another example for GPCR is the α2-adrenergic receptor and opioid drug receptors. These receptors might release closely related $G_{\alpha i}$ as a GTP complex. $G_{\alpha i}$ inhibits adenylate cyclase, and opposes the effect of Gαs. Gαs is a slow GTPase, so it hydrolyses the GTP and convert it into its inactive GDP state. This results into the deattachment of GDP from adenylate cyclase in inactive form. Gαs-GDP forms heterotrimeric complex with GPCR till fresh ligand arrive and bind to receptor. Adenylate cyclase do not possess the GTPase accelerator protein (GAP) activity and Gαs decline remains. GTPase activity controls the time period in which Gαs is active. $G_{\alpha i}$ also hydrolyses GTP very slowly; though some of its activation targets like phospholipase Cγ do act as GAPs (Table 9.2). Enzyme 3,' 5'-cyclic nucleotide phosphodiesterase break the Cyclic AMP. Caffeine represses the enzyme and helps in the release of cyclic AMP which induces the release of glucose sustaining a high catabolic rate longer than usual. Phosphodiesterase 5 isozyme leads cyclic GMP for hydrolysis. The cyclic GMP increases blood flow and relaxes vascular smooth muscle. Sildenafil (Viagra) is a particular specific inhibitor of the phosphodiesterase 5 isozyme.

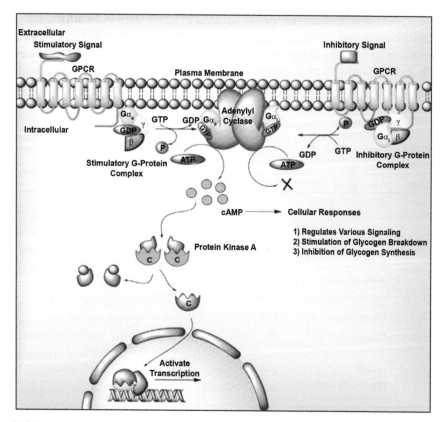

FIGURE 9.2 Schematic representation of stimulatory and inhibitory ligand binding to GPCR and their respective response in the cell. In both G-proteins, $G_{\beta\gamma}$ subunit is similar. Only G_{α} or respective G- protein subunits differ.

Phosphodiesterase inhibition maintains high cyclic GMP levels for a longer period. $G_{\beta\gamma}$ pair released from $G_{\alpha s}$-GTP regulates certain processes, such as stimulation of some specific enzymes (tyrosine kinases and phospholipase Cγ) (Figure 9.3).

9.2.2 PHOSPHATIDYLINOSITOL (PI) SIGNAL PATHWAY

PI, a lipid present on membrane involved in signal cascade triggered by certain hormones. Kinases facilitate stepwise transfer of inorganic phosphate of ATP to yield phosphatidylinositol-4,5- bisphosphate (PIP2).

Phospholipase C cleaves PIP2. G-protein (G_q) activates a particular phospholipase C. When a ligand binds to a specific GPCR (receptor) it gets activated and exchanges GTP for GDP. After that, G_q-GTP activates Phospholipase C. The activation requires Ca^{++} which interacts with oppositely charged phosphate groups of the phosphorylated inositol at the active site.

TABLE 9.2 Major Classes of Trimeric Gα-Proteins and Their Effectors in Mammalian Systems

Class	Associated effector	Second Messenger	Receptor Examples
Gsα	Adenylyl cyclase	cAMP ↑	β-Adrenergic (epinephrine) receptor, Receptors for glucagon serotonin and vasopressin
Giα	Adenylyl cyclase K⁺ channel (GβΥ activates effector)	cAMP ↓ Deviation in membrane potential	α1-Adrenergic receptor mAChRs (Muscarinic acetylcholine receptor)
Go1fα	Adenylyl cyclase	cAMP ↑	Odorant receptors (innose)
Gqα	Phospholipase C	IP-3, DAG ↑	α2-Adrenergic receptor
Goα	Phospholipase C	IP-3, DAG ↑	Acetylcholine receptor (AChRs) in endothelial cells
Gtα	cGMP phosphodiesterase	cGMP ↓	Rhodopsin (light receptor) present in rod cells (inside retina as a light receptor)

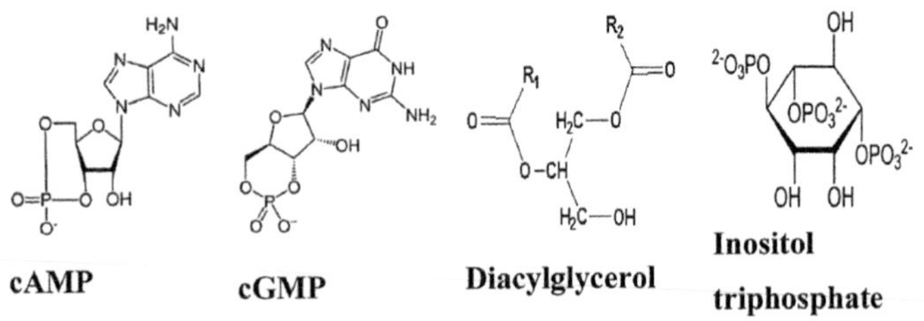

FIGURE 9.3 General intracellular second messengers convoluted in signal transduction.

Phospholipase C cuts PIP2 into two-second messengers: diacylglycerol (DG) and inositol- 1,4,5-trisphosphate (IP-3) (Figure 9.3). DG in the association through Ca^{++} stimulates Protein Kinase C that phosphorylates several cellular proteins and thereby changes their activity. IP-3 (inositol-1,4,5- trisphosphate) activates calcium ion channels in ER (endoplasmic reticulum) resulting into discharge of ions in the cytosol. The phenomenon is carried out by calmodulin protein, a target of protein kinase (Figure 9.4).

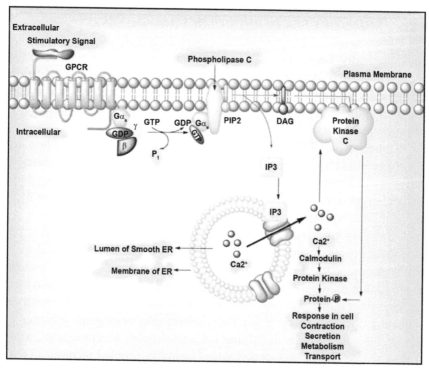

FIGURE 9.4 Cytosolic calcium ion level increase through phosphatidylinositol signaling pathway.

Degradation of IP3 results into Ca^{++}-adenosine triphosphatase (ATPase) pumps mediated efflux of Ca^{++} in ER lumen from the cytosol. IP-3 (inositol-1,4,5-trisphosphate) converted into inositol by sequential dephosphorylation catalyzed by enzymes used in the synthesis of PI. In another reaction, IP3 may get phosphorylated by specific kinases, converting it into IP-4, IP-5, or IP-6. These products have role in signaling for example in some cells IP-4 (inositol-1,3,4,5-tetraphosphate) excites Ca^{++} admittance by membrane Ca^{++} channel stimulation (Figure 9.4). Kinases involve in the conversion of PI to PIP2, facilitate phosphate transfer from ATP to OH groups at 4th and 5th position in the inositol ring.

PI 3-Kinases catalyze phosphorylation of PI (3, 4, and 5) triphosphate at the three places in inositol ring. PI-3-P, PI-3, 4-P2, PI-3, 4, 5-P3, and PI-4, 5-P2 have signaling roles.

9.2.3   ROLES OF GPCR IN PHYSIOLOGY AND DISEASES

Physiologically important GPCR mediates hormones effect especially peptide hormones. An additional type of GPCR is found in neurotransmission process such as receptors for norepinephrine, acetylcholine, dopamine, serotonin, glutamate, as well as rest of the lipids and peptides that function as neuromodulators.

A number of GPCR are involved in hormonal signaling. They are documented as important pathological entities in many endocrine diseases such as autosomal dominant hypocalcemia (ADH), Hirschsprung's disease, cryptorchidism, etc. Mutations in Frizzled receptors are involved in Familial exudative vitreoretinopathy (FEVR). The mutations are known to stop hormonal response. Monogenic mutations are found in the rhodopsin genetic disorder retinitis pigmentosa (RP) patients. Rhodopsin family A comprises the highest number of GPCR superfamily members. Nephrogenic or renal diabetes insipidus (NDI) is a consequence of the vasopressin failure. Vassopressin maintains the water reabsorption in collecting duct of kidney. It is a most contrasting monogenic disorder of GPCR. Genetic disorder involving calcium-sensing receptor is example of monogenic disease associated with G-protein coupled receptor (Tfelt-Hansen et al., 2005). The receptor is present in many tissues involved in calcium homeostasis. Polymorphisms in GPCR gene is also reported and is associated with human diseases. For example, Dopamine receptor 2 and Dopamine receptor 3 are constructively associated with depression, anxiety, and schizophrenia respectively.

9.2.4   WNT SIGNALING

The Wnt/ß-Catenin pathway controls cell destiny decisions as well as pluripotency of stem cells during developmental process. These developmental cascades assimilate signals from several other pathways, including bone morphogenetic protein (BMP), retinoic Acid (RA), Growth Factor of fibroblast (FGF) and transforming growth factor ß (TGF-ß). Frizzled receptor bind with a secreted glycoprotein (Wnt ligand) and produce a larger complex with LRP5/6 proteins (Figure 9.5).

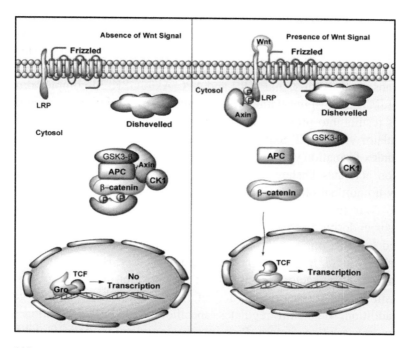

(A)

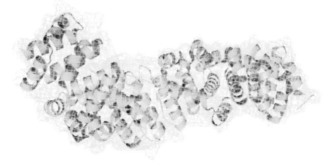

(B)

FIGURE 9.5 (A) Signal transduction through Wnt pathway, (B) cartoon structure of β-catenin (PDB: 1JDH) protein which moves into the nucleus in the presence of Wnt signal and transcribes targeted genes.

E3 ubiquitin-protein ligase, zinc, and ring finger 3 (ZNRF3) and its homolog ring finger 43 (RNF43) are involved in ubiquitination of Frizzled receptor. R-spondin prevents ZNRF3 and RNF43 activity by binding with Lipoprotein receptor-related Proteins 5/6 (LRP5/6). Thus, R-spondins increased the susceptibility of cells for Wnt ligands. Wnt receptor activation triggers the dislocation of multipurpose kinase glycogen synthase kinase-3 ß (GSK-3ß) from a regulatory APC/Axin/GSK-3ß-complex (Figure 9.5). In the absence of Wnt-signal, CK1 and APC/Axin/GSK-3ß-complex provide coordinated phosphorylation of ß-catenin and transcriptional co-regulator. This results into proteasomal degradation of β-catenin of transcriptional regulator via ß-TrCP/Skp pathway. The LRP5/6 co- receptor involves in complex formation with Wnt-associated Frizzled receptor and ligand. This action activates Disheveled (Dvl) protein by stepwise phosphorylation, poly-ubiquitination, and polymerization. This relocates GSK-3ß from APC/Axin (mechanism not known) by the process of substrate trapping and/or endosome sequestration. Stabilized form of ß-catenin is translo-cated towards nucleus through Rac1 and several another factors where ß-catenin binds to LEF/TCF transcription factors. Binding of ß-catenin to transcription factors results in the transposition of co-repressors and enlisting of supplementary co-activators to the Wnt-specific target genes. In addition, ß-catenin regulates specific targets in cooperation with other transcription factors. Prominently, investigators have found point mutations in the ß-catenin present in human tumors. It preventsGSK-3ß phosphorylation and leads to its anomalous accumulation. Mutations in genes like R-spondin, Axin, APC, and E-cadherin has been reported in tumor samples, emphasizing the deregulation of Wnt pathway in cancer. Wnt signaling also promote aggregation of other transcriptional regulators (TAZ and Snail-1) in the nuclei of cancer cells.

9.2.5   *NOTCH SIGNALING PATHWAY*

Notch signaling pathway is frequently involved in cell-type specification, organogenesis, and development. Notch genes got their name after a pheno-type of mutant Drosophila which has notched wing. The gene encodes highly sustained cell surface notch receptors (NR). The mammalian NRs (hNotch1, hNotch2, hNotch3, and hNotch4) are enormous, transmembrane proteins of type 1 category contain numerous structural motifs (Figure 9.6).

Furin-like convertases cleaves the NR in trans-Golgi network, yielding two subunits of the complete NR.

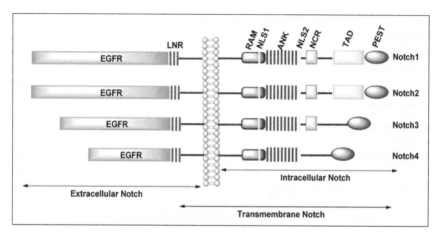

FIGURE 9.6 Structure of different human notch receptors involved in notch signaling.

The extracellular notch (ECN) is mainly comprised of a ligand-associated domain containing three LIN12/Notch repeats and tandem EGF like Repeats (EGFR). These are found to restrain unsuitable and ligand-independent receptor activation. Membrane-bound Notch subunit contains small extracellular domain, (the single transmembrane domain) in addition to a bulky intracellular domain. Intracellular domain has seven repeats of cdc10/ankyrin-like repeats, a RAM sequence, a C-terminal PEST (P for proline, E for glutamic acid S for serine and T for threonine) sequence and two nuclear localization signals (NLS). In addition, mammalian Notch1, Notch2, and Notch3 include cytokine response (NCR) sections and Notch 1, Notch 2 has a C-terminal domain for transcriptional activation (TAD). Notch signaling initiates with the NRs present on cell exterior.

Ligand binding triggers two subsequent proteolytic cleavages that release the active intracellular part of the Notch receptor (ICN) (Figure 9.7). If Notch's extracellular domain binds a ligand, an ADAM-family (A Disintegrin and A Metalloprotease) metalloprotease named ADAM10 cuts the notch protein fairly from the exterior of membrane. Structure of ADAM protein is shown in Figure 9.8. This cleavage is usually known as S2 cleavage. After S2 cleavage reaction, gamma- secretase (also associated with Alzheimer's disease) cuts the rest part of the NR just inside the interior sheet of cell membrane. Structure of gamma secretase is shown in Figure 9.8. After this cleavage notch intracellular domain (NICD), translocate inside the nucleus. In nucleus, NICD controls gene expression by facilitating the transcription factors like CSL. In the beginning, it was thought that CSL protein suppress Notch targeted transcription. However, it is found that CSL modulates its activity from a transcription repressor to activator after binding with the intracellular domain of complex.

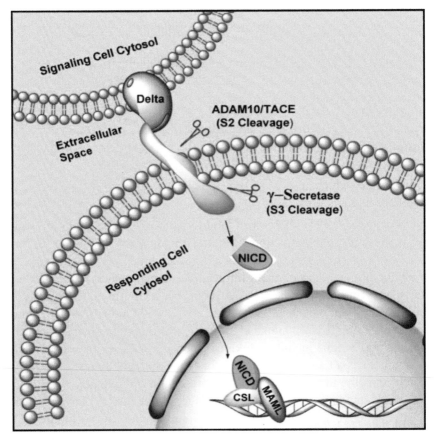

FIGURE 9.7 Schematic notch signaling pathway viewing a simple overview of the basic mechanisms for the transcription target genes induction.

There are some other proteins involved in intracellular portion of Notch signaling pathway. ICN goes inside the nucleus, to act with some regulatory transcription factors. The only transcription factor identified to bind ICN (intracellular domain of notch) is CSL. In the absence of ICN, CSL acts as a transcription repressor in association with histone deacetylases (HDAC) and co-repressors (Kao et al., 1998). Association of ICN with CSL disturbs complex and acts as a corepressor. It recruits coactivators and promotes transcription of CSL-binding elements.

The transcriptional activation complex of notch or enhanceosome is a quite large protein complex. This complex holds ICN, CSL (CBF1, Suppressor of Hairless or J kappa-recombination signal-binding protein and Lag-1) and Mastermind-like (MAML) polypeptides. Reconstitution of transcription (*In-vitro*) from CSL-mediated promoter elements on the chromatin template involves Mastermind-like protein (MAML1), ICN, and CSL (Fryer et al., 2002). Binding of MAMLs (glutamine-rich nuclear proteins) to Ankyrin repeat (a section of Notch) forms constant ternary complex on DNA with CSL, ICN, and recruits CBP/p300 (histone acetylase) (Fryer et al., 2002). MAML1 intensifies CSL reporter gene activation whereas truncated MAML1 acts as a dominant- negative inhibitor of CSL activation. In addition, other possible components of the complex includes SKIP protein which interacts with CSL-ICN co-repressor complex and facilitates its transformation into an activator complex.

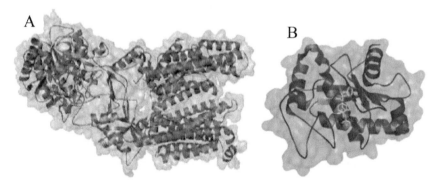

FIGURE 9.8 (A) Cartoon structure of human ϒ-secretase (PDB: 5A63) is responsible for S3 cleavage, i.e., from the inside of the membrane, (B) cartoon structure of ADAM-8 metalloproteinase domain (PDB: 4DD8) with bound batimastat (lemon color). Its homolog ADAM10 cleaves the Notch proteins from the exterior of the membrane, i.e., S2 cleavage.

Mutations in Notch signaling pathway proteins may cause developmental phenotypic abnormalities which affect the function of heart, kidney, liver, eye, skeleton, face, and vasculature. Various Notch linked disorders are autosomal dominant, autosomal recessive (spondylocostal dysostosis-mutations in Delta-like-3 ligand), Alagille syndrome (mutations in both Jagged1 ligand and NOTCH2 receptor). NOTCH2 mutation is associated with Hajdu-Cheney syndrome, and a dominant ailment triggers osteoporosis, focal bone devastation, renal cysts, and craniofacial morphology. NOTCH1 receptor mutation is associated with various types of cardiac disease. NOTCH3 mutation instigates the dominant adult-onset disorder CADASIL syndrome.

9.2.6 SONIC HEDGEHOG (SHH) PATHWAY

Hedgehog (HH) pathway performs essential role in developmental process. The hedgehog family comprises of three proteins namely SHH, desert hedgehog (DHH), and Indian hedgehog (IHH). Sonic HH regulates vertebrate organogenesis process such as development of brain and development of limb digit. SHH; a morphogen molecule circulate in cells of propagating embryo in a concentration- dependent manner. The concentration gradient differentially regulates the developing embryo. Hedgehog genes were first observed by Edward Lewis along with Christiane Nusslein-Volhard and Eric Wieschaus in 1978. They were awarded by the Nobel Prize (1995) for the recognition of genes that regulate segmentation pattern in *Drosophila melanogaster* embryo (Figure 9.9). Situated towards embryo's posterior part there is a structure called, ZPA (zone of polarizing activity), secretes SHH in limb bud of embryo (Figure 9.10).

SHH (45 kDa) synthesis is carried out in rough endoplasmic reticulum (RER). After auto-processing, it produces an N-terminal signaling domain (20 kDa) and a C-terminal domain (25KDa) with unknown signaling function. SHH conjoins with a transmembrane protein commonly known as patched (PTCH). PTCH inhibits over-appearance and action of the seven membranes panning smoothened (SMO) receptor in the absence of hedgehog protein. Extracellular hedgehog is known to associate and inhibit PTCH protein.

This event allows smoothened to gather and prevent the ubiquitin-mediated proteolytic cleavage of the transcription factor, *Cubitus inter-ruptus* (Ci). Thus, the association of HH and PTCH allows, Ci protein to get into the nucleus and function as transcription factor inside the nucleus (Figure 9.10). In the absence HH, Ci associates with costal-2 (cos2) kinesin-like protein to form complex. The complex induces the protea-somal degradation of Ci protein in CiR fragments. CiR has capability to suppress transcription of specific target gene. HH-activated Patched induces the gathering of Ci protein in cytoplasm thus reduces CiR level and permits transcription of certain specific genes (e.g., decapentaplegic). Expression of genes regulated by other HH requires loss of CiR along with the optimistic action of whole Ci acting like a transcriptional activator.

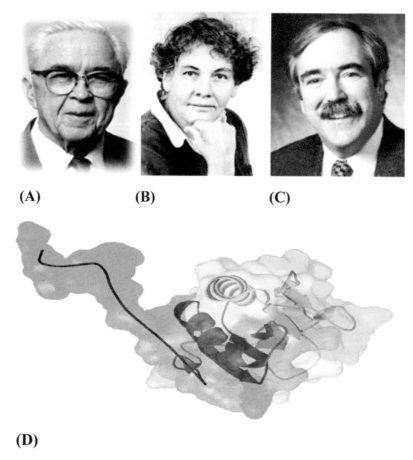

(A) **(B)** **(C)**

(D)

FIGURE 9.9 A) Edward B. Lewis, B) Christiane Nüsslein Volhard, C) Eric F. Wieschaus were awarded Nobel Prize (1995) for identifying the genes that regulate the pattern of segmentation in *Drosophila melanogaster* embryos D) human sonic hedgehog (PDB: 3M1N) N-terminal domain.

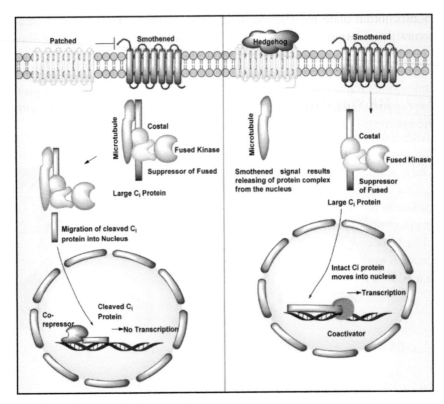

FIGURE 9.10　Sonic hedgehog pathway.

Hedgehog pathway activation also yields increased expression of "Snail" protein, reduction in tight junctions and cadherin. Hedgehog signaling also seems to be an important regulator of angiogenesis (formation of new blood vessels) and metastasis. Hedgehog pathway activation yields an increased expression of cyclin D1/B1, angiopoietin-1/2 and anti-apoptotic genes (i.e., BCL-2 Family) and a declined expression of apoptotic genes (e.g., Fas). Signaling also plays an important role in gene expression regulation in mature stem cells. It controls the preservation and renewal of adult tissues. It is also been found that hedgehog pathway is involved in the onset of some cancers. Holoprosencephaly results due to damage in the ventral midline (i.e., Midline mesectoderm) caused by alterations inhuman SHH gene. Formation of some special category of cancer tumor has been linked with SHH pathway. SHH is required for the division of unilateral eye field to two bilateral fields. SHH produced by prechordal plate inhibit Pax-6 protein. This repression facilitates the separation of two eye field. Mutation in SHH gene was found in cyclopia, a phenotype with one eye in the midpoint of the forehead.

9.2.7 JAK-STAT PATHWAY

JAKS (Janus kinase) and STATS (signal transducer and activator of transcription proteins) are important modules for several cytokine receptor systems which regulate survival, differentiation, growth, and pathogen resistance. In this pathway, signal binding encourages the dimerization of receptor leading to the activation of allied JAK proteins. JAK proteins phosphorylate receptors and themselves also. Protein containing SH2 domain, for example, STAT3 docks with the phosphorylated site present on receptor and JAKS. Once Phosphorylated Stat protein get dimerize and transported in the nucleus to modulate transcription of the targeted gene (Figure 9.11).

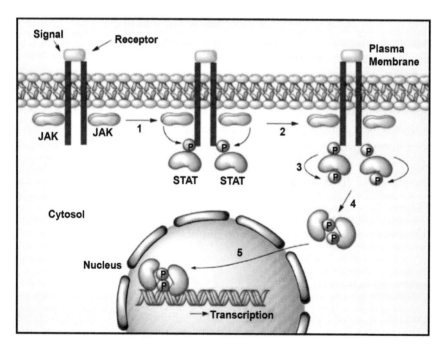

FIGURE 9.11   JAK-STAT pathway showing ligand-receptor binding and activation of JAK kinases. Phosphorylation of STAT proteins by JAK kinase results into dimerization, transportation inside the nucleus and transcription of the specific target genes.

SOCS protein (suppressor of cytokine signaling) diminishes receptor signaling cascade through heterologous/homologous feedback regulation. JAK/STAT pathway arbitrates the property of cytokines, thrombopoietin (TPO), erythropoietin (EPO), and G-CSF. They act as protein drug for the treatment of thrombocytopenia, anemia, and neutropenia respectively. This pathway also facilitates interferon signaling utilized as anti-proliferative and antiviral agents. It has been reported that unregulated expression of cytokine signaling is involved in cancer pathology. Anomalous IL-6 signaling aids in the pathological process of inflammation, autoimmune diseases along with cancer i.e., multiple myeloma and prostate cancer (PCa). Inhibitors of JAK-STAT pathway are nowadays being inspected in model organisms of multiple myeloma. Constitutive active STAT3 in tumors established its oncogenic role. In glioblastoma cells, overexpressed EGFR defiance to EGFR tyrosine kinase inhibitors encouraged by JAK2-EGFR binding.

Most of the human hematological malignancies are the result of JAK mutations. Studies revealed a distinctive somatic mutation in pseudokinase domain of JAK 2 (V617F) frequently aroused in essential thrombocythemia, idiopathic myelofibrosis, and polycythemia vera. This type of mutations cause pathological activation of JAK2 allied with EPO, TPO, and G-CSF receptors. These proteins govern erythroid, megakaryocytic, and granulocytic differentiation and proliferation. Somatically attained gain-of-function mutations in JAK1 are established in T-ALL (T-cell acute lymphoblastic leukemia) in adult. These types of mutations in JAKs have been found present in ALL (pediatric acute lymphoblastic leukemia). Moreover, mutations in JAK2 have been spotted nearby pseudokinase domain R683 (R683G/DIREED) in childhood Down syndrome, B-acute lymphoblastic leukemia, and pediatric B-acute lymphoblastic leukemia. It has been found that STAT3/5 is constantly stimulated by tyrosine kinases different from JAKS in various solid tumors.

9.2.8 NF-κB PATHWAY

It was discovered by Dr. Ranjan Sen in the laboratory of David Baltimore. They proved its interaction with the eleven Bp sequence in antibodies (immunoglobin) light chain enhancer of B cells. NF-kB bears a substantial role in cellular growth, inflammatory reactions, apoptosis, and several diseases such as cancer, arthritis, asthma, etc. NF-κB belongs to an extensively encountered group of transcription factors like Rel protein dimers. The Rel group member comprises RHR (Rel Homology Region), a structurally analogous motif. It is accountable for DNA binding. Other Rel proteins dimerize adjacent to DNA. RHR interacts with several inhibitor proteins in the cytoplasm. Due to similarity in their structure, they could be classified into two classes based on the C-terminal sequence of the RHR. In the first class, member of p105 class is the precursor of NF-kB's monomers. After activation, the repeated chain gets cleaved and P105 becomes p50. The second class of Rel protein, p65 (RelA) encloses a transcription activator section on the carboxyl-terminus of RHR. The p65 subunit of NF-kB has synergy as well as an activation domain indispensable for the enhanceosome's success (Figure 9.12).

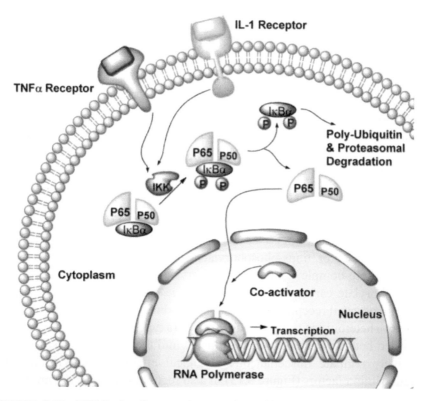

FIGURE 9.12   NF-kB signaling protein executing a biochemical signaling pathway resulting in the transcription of marker genes.

Homodimer of the Rel proteins also occurs but cannot induce transcription. p50 and p65 dimerize nearby a ten base pair region called as a kB site, creating the NF-kB transcription factor. The binding sequence is 5'-GGGRNYYYCC-3' where Y denotes pyrimidine, R signifies a purine, and N is an unspecified base. I-kB kinase can be stimulated by different stimulus such as tumor necrosis factor (TNF)-an IL-10, toll-like receptor activation, virus infection, ionizing radiation, and binding of pro- inflammatory cytokines. Binding of specific signal with the particular receptor activates I-kB kinase. I-kB kinase then phosphorylates the I-kB resulting into the detachment with IκBa from NF-κB (p65/p50). E3 ubiquitin ligase binds to phosphorylated site and poly-ubiquitinates IκBa triggering its immediate degradation by proteasome. Removal of IκBa unmasks the nuclear-localization signal in NF-kB. In the nucleus, NF-kB executes the transcription of various target genes (Figure 9.12).

9.2.9   TGF-SS SIGNALING

TGF-ß superfamily signaling is critical in the maintenance of cell differentiation, and growth along with the development in use of cells in widespread biological systems. Ligands of TGF-ß (Transforming Growth Factor-ß) receptor are either homodimers or heterodimers with a set of seven cysteine residues. There are two varieties of cell surface receptor playing a fundamental part in TGF-ß signaling i.e., Type I, Type II. This receptor belongs to serine-threonine kinases family. All Type I receptor have a TTSGSGSG motif called as GS domain. Signaling by TGF-ß involves binding of ligand and establishment of a ternary holo complex that brings together both the Type I and Type II receptors. Activated type II kinase receptor, trans-phosphorylates the type 1 receptor at GS domain in the holo-complex. These phosphorylated Type I receptors finally phosphorylate carboxy-terminal of the mad and activate Smad proteins. Phosphorylated receptor regulates the mad dimerization/trimerization and forms a heteromeric complex with Smad 4 (a type of co-Smad). This newly formed heteromeric complex translocate into the nucleus. Smad proteins actively regulate miscellaneous biological effects in association with transcription factors (Figure 9.13). It also regulates cell-specific inflection of transcription. Apart from it, I-Smads negatively regulate both active in/ TGF-ß and BMP signaling by competing with R-Smads for the receptors (negative feedback loop). I-Smads also targets co-smad for degradation.

Smads are commonly known as cytoplasmic signaling proteins. Up to date, receptor-regulated smad (R-Smad/Smad 1, 2, 3, 5, and 8), inhibitory smad (I-Smad/Smad 6,7) and co-mediator smad (co-Smad) are known. Smad 2/3 are involved in TGF-ß/activin pathway and Smad1/5/9 is involved in BMP pathway. The firmness of TGF-ß family receptor and smad proteins are controlled by Smurf E3 ubiquitin ligase. TGF-ß pathways are also regulated by MAPK signaling pathway in the biological system. In certain circumstances, TGF-ß signaling can also distress by ERK, SAPK/JNK, and p 38 MAP kinase pathways. RhoA GTPase triggers downstream proteins such as ROCK for the rapid reorientation of cytoskeleton allied with cell growth/spreading regulation and cytokinesis. Modulation of PKC and c-Abl leads to TGF-ß activation by Cdc42/Rac.

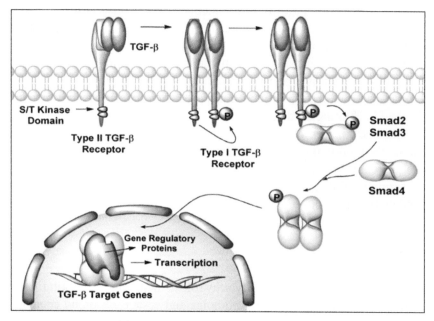

FIGURE 9.13 TGF-β signaling pathway.

9.3 RECEPTOR TRAFFICKING

Regulation of intracellular transportation such as tethering, budding, a fusion of vesicles/organelle, etc. needs Rab GTPases. Rab GTPases also regulate GPCR trafficking considered as a comprehensive process enabling the trafficking of receptors between compartments of the intracellular membrane. Recently, studies showed a very gradually undoubted sign that trafficking of receptor unsettles the signal transduction of cargo proteins. These vesicular cargo signaling can transmogrify Rab GTPase monitored intracellular transportation processes.

This type of results explored the correlation of Rabs with other Rab proteins as well as regulatory molecules to facilitate protein trafficking. It demonstrates the dynamic role of these transport molecules in disease pathology. Esseltine and Ferguson (2013) discovered exocytotic small G protein Rab8. It plays a unique role in the signal transfer of metabotropic glutamate receptor-1 and intracellular trafficking. Another study showed that Rab GTPase (known as Rab11) also involved in the pathology of Huntington disease. Rab11-dependent vesicle formation is diminished in fibroblasts of Huntington disease patients. Adult mouse brainexpressingRab11 dominant-negative protein, showed a parallel neurodegenerative phenotype in mutant mouse models of Huntington disease. Association of Rab 8, Rab11, and myosin Vb regulates trafficking (Esseltine and Ferguson, 2013). It is promising that Rab8 and Rab11 work in association to modulate mGluR1a/5 protein transport under standard conditions and changed in Huntington disease.

Tyrosine kinases associated receptors such as activated EGFR are internalized through endocytosis (Haigler et al., 1979) led to the assumption that conveyance of ligand to the nucleus where it openly affects gene transcription. Rab5, Hrs, and Cbl are the three main model proteins that regulate tyrosine kinase receptor movement.

9.3.1   RAB5

Rab protein GTPases play a noteworthy role in membrane trafficking. They control the assembly of SNARE complex and vesicle fusion by engaging tethering factors e.g., EEA1 (early endosomal autoantigen 1), that is called up by Rab5. Rab5 is limited only to the early endosome. It binds with numerous effector molecules and modulates both motility and endosome fusion. Endocytic clathrin- coated vesicle formation is also regulated by Rab5. It is found that stimulated endocytosis is linked to Rab5 activation and is Rab5 dependent (Barbieri et al., 2000). Accumulation of EGF into serum-starved cells resulted into GTP packing of Rab5. These events comprise

PI3K, protein kinase B (PKB) because wortmannin and dominant-negative PKB/Akt appearance block the activation of endocytosis by EGF. Another study showed the combination of Rab5 with two PI 3-kinases (hVPS34 and p110ß) and a Rab5-GTP. The hVPS34 enzyme produces PI 3-phosphate which in turn conjoins with Rab5 to introduce EEA1 in endosomes. Here it takes part in endosome fusion events (Simonsen et al., 1998). The communication between p110ß and Rab5 indicates that Rab5 plays role in the localized construction of phosphatidylinositol-(3,4)-bisphosphate and phosphatidylinositol-(3,4,5)- trisphosphate and resulting in the action of downstream effectors (e.g., PKB).

9.3.2 HRS

Hepatocyte growth factor controlled tyrosine kinase substrate (Hrs) is also called as Hgs (hepatocyte growth factor is an endosome-localized protein). It is a protuberant goal for phosphorylation in downstream of a range of receptors including EGFR, Met, GM-CSF, and interleukin 2. It also shows the connection of FYVE-finger motif and a VHS domain (an acronym for VPS 27, Hrs, and STAM) with the yeast protein Vps27p. Vps27p is in the E class Vps mutant showed malfunctioning in trafficking (from the sorting endosome to the vacuole). Internalization of EGFR in Hrs-comprising endosomes is compulsory for EGF related Hrs phosphorylation (Urbé et al., 2000). These phosphorylation events needed PI 3-kinase activity as showed by its sensitivity to wortmannin. Once phosphorylated, Hrs appears to be free from endosomal membrane and reaches to cytosol (Urbé et al., 2000).

9.3.3 CBL

Caenorhabditis elegans vulval development directed the documentation of Cbl homolog, SLI-1 (suppressor of lineage (SLI) defects). It acts as a negative modulator of tyrosine kinase receptor LET- 23 and downstream signaling. Cbl works as a proteasomal E3 ubiquitin ligase involving both E2 ubiquitin-conjugating enzymes and phosphorylated receptors by binding with RING-finger domain and tyrosine-kinase binding (TKB) domain respectively. This brings receptor polyubiquitination which specifies proteasomal degradation and thus upregulate the transport to late- endosome/lysosomal compartments (Bonifacino and Weissman, 1998).

KEYWORDS

- autosomal dominant hypocalcemia
- bone morphogenetic protein
- *Cubitus interruptus*
- desert hedgehog
- erythropoietin
- extracellular notch

REFERENCES

Barbieri, M. A., Roberts, R. L., Gumusboga, A., Highfield, H., Alvarez-Dominguez, C., Wells, A., & Stahl, P. D., (2000). Epidermal growth factor and membrane trafficking: EGF receptor activation of endocytosis requires Rab5a. *J. Cell Biol.*, *151*(3), 539–550.

Bonifacino, J. S., & Weissman, A. M., (1998). Ubiquitin and the control of protein fate in the secretory and endocytic pathways. *Annu. Rev. Cell Dev. Biol.*, *14*(1), 19–57.

Esseltine, J. L., & Ferguson, S. S., (2013). Regulation of G protein-coupled receptor trafficking and signaling by Rab GTPases. *Small GTPases*, *4*(2), 132–135.

Fryer, C. J., Lamar, E., Turbachova, I., Kintner, C., & Jones, K. A., (2002). Mastermind mediates chromatin-specific transcription and turnover of the notch enhancer complex. *Genes Dev.*, *16*(11), 1397–1411.

Haigler, H. T., Mckanna, J. A., & Cohen, S., (1979). Direct visualization of the binding and internalization of a ferritin conjugate of epidermal growth factor in human carcinoma cells A- 431. *J. Cell Biol.*, *81*(2), 382–395.

Kao, H. Y., Ordentlich, P., Koyano-Nakagawa, N., Tang, Z., Downes, M., Kintner, C. R., Evans, R. M., & Kadesch, T., (1998). A histone deacetylase corepressor complex regulates the notch signal transduction pathway. *Genes Dev.*, *12*(15), 2269–2277.

Simonsen, A., Lippe, R., Christoforidis, S., Gaullier, J. M., Brech, A., Callaghan, J., Toh, B. H., Murphy, C., Zerial, M., & Stenmark, H., (1998). EEA1 links PI (3) K function to Rab5 regulation of endosome fusion. *Nature*, *394*(6692), 494–498.

Tfelt-Hansen, J., Ferreira, A., Yano, S., Kanuparthi, D., Romero, J. R., Brown, E. M., & Chattopadhyay, N., (2005). Calcium-sensing receptor activation induces nitric oxide production in H-500 Leydig cancer cells. *Am. J. Physiol. Endocrinol. Metab.*, *288*(6), E1206– E1213.

Urbé, S., Mills, I. G., Stenmark, H., Kitamura, N., & Clague, M. J., (2000). Endosomal localization and receptor dynamics determine tyrosine phosphorylation of hepatocyte growth factor- regulated tyrosine kinase substrate. *Mol. Cell. Biol.*, *20*(20), 7685–7692.

CHAPTER 10

Human Disease Drug Therapy and Drug Discovery

PREM PRAKASH KUSHWAHA,[1] REBATI MALIK,[1]
P. SESHU VARDHAN,[2] SHIV GOVIND RAWAT,[3] ATUL KUMAR SINGH,[1]
and SHASHANK KUMAR[1]

[1]*School of Basic and Applied Sciences, Department of Biochemistry, Central University of Punjab, Bathinda, Punjab–151001, India, Tel.: +91 9335647413, E-mail: shashankbiochemau@gmail.com (S. Kumar)*

[2]*School of Biotechnology, Jawaharlal Nehru Technological University, Kakinada–500085, Telangana, India*

[3]*Department of Zoology, Banaras Hindu University, Varanasi–221005, Uttar Pradesh, India*

ABSTRACT

Intracellular and extracellular factors are well known for maintaining body homeostasis in any organism. An anomalous condition which undesirably regulates internal biomolecular pathways, originates an ailment. Several hormonal therapies, epigenetic modification, targeting various receptors such as HER2, EGFR, and potassium channels are the major targets to reduce the effects of a disease. Nowadays, monoclonal antibodies, nanobodies, and biosimilar drugs are creating major attraction against the numerous disease treatments.

10.1 INTRODUCTION: HUMAN DISEASE AND DRUG THERAPY

Drug development is an extremely regulated and complex procedure, comprise of an academic research institution, the pharmaceutical industry, and government agencies, Current research in the pathogenesis of various diseases have produced recurrent drug targets. This results into gathering of several small molecules as clinical trial drugs.

10.1.1 CANCER: A DEVASTATING DISEASE

Cancer is a serious health issue in developed states. It is related with the aging and lifestyle of the population. Early diagnosis, general access to medical services and advancement in treatment brought a noteworthy change in cancer survival. Outcome of disease treatment has extensively improved over the last four decades. The era when both surgery and radio-therapy were the only most prominent approach to treat the tumor growth has been ended (Baskar et al., 2012). A scenario where the molecular aspect of tumors seems to be the keystone of any therapy is now growing. Here we will discuss various strategies for cancer treatment.

10.1.1.1 ENDOCRINE THERAPY

Hormonal dependence of breast cancer has been used to fight the disease. Several approaches applied, ranging from ovarian ablation in pre-menopausal women to the recently introduced aromatase inhibitors (AIs) including specific estrogen receptor modulators (SERMs). Tamoxifen, estrogen is a dual agonist-antagonist that interacts to estrogen receptor obstructs the binding of co-activators; thereby inhibit the ER-induced transcriptional activity (Shiau et al., 1998). Recently, tamoxifen showed promising anticancer activity against postmenopausal patients. Third generation AIs such as anastrozole letrozole and exemestane are also used in therapy. Taking into explanation the favorable toxicity outline of the AIs, they have rapidly become the first-line of treatment (Johnston and Dowsett, 2003). Each one of the three third-generation AIs has shown lengthy disease-free survival rates. Endocrine management continues to base systemic methods for the treatment of prostate cancer (PCa).

10.1.1.2 ANTI HER-2 TARGETED THERAPY

Human epidermal growth factor (EGF) receptor 2/neu (HER-2/neu) belongs to EGFR (epidermal growth factor receptor) family. They have tyrosine kinase activity which facilitates cell growth, differentiation, and survival (Yarden and Sliwkowski, 2001). Amplification or overexpression of HER-2/neu gene or both has been found in breast cancer. ErbB2 receptor is an extracellular target, for precise anticancer treatment. Several studies have been reported to target the extracellular domain of ErbB2 with monoclonal antibody (mAb) trastuzumab. Trastuzumab stimulates ErbB2 internalization and deprivation resulting into the improvement of antibody-dependent cytotoxic reactions. Higher ErbB2 level expression is known to inhibit the development of ErbB2 positive human tumor xenografts. Other than trastuzumab, a new molecule Lapatinib (Tykerb) got approval for the cure of HER2-positive metastatic breast cancer. Lapatinib is a small molecule that inhibits the intracellular tyrosine-kinase activity of EGFR and Her-2 receptors delaying its ATP-binding function. Combined chemotherapy of trastuzumab and lapatinib showed better progression-free survival rates in cancer treatment (Geyer et al., 2006).

10.1.1.3 ANTI EGFR TARGETED THERAPY

The EGFR; (ErbB-1) belongs to the closely related ErbB receptor family. It serves as a target for mAbs and small molecule inhibitors. Cetuximab, an anticancer drug commercially known as Erbitux. It is a chimeric mouse/human mAb raised against the extracellular portion of EGFR. Cetuximab is FDA approved and used in the combination chemotherapy of metastatic colorectal cancer.

Panitumumab (a human IgG2 antibody) is another potential molecule used in the management in metastatic colorectal cancer chemotherapy (Van Cutsem et al., 2007). Some tyrosine kinase inhibitors of EGFR (gefitinib and erlotinib) have secured FDA approval for anti-tumor activity. Molecular changes in EGFR (either mutations or in-frame deletions) are allied with lung cancer treatment. These EGFR mutations confer a healthy response to the ligand and a consistent receptor inhibition potential of gefitinib. Cetuximab showed a clinical advantage against colon cancer as it increases EGFR expression independent of mutation in EGFR (Sartore-Bianchi et al., 2007). In malignant glioma, sensitivity is strictly associated with deletions inside EGFR ectodomain.

10.1.1.4 EPIGENOMIC TARGET

Epigenetic alterations are reversible. In cultured cancer cell lines, some molecules have potential to re-express the genes that are silenced by methylation. These molecules are known as DNA- demethylating molecules. Low doses of these molecules have demonstrated a noteworthy antitumoral potential in patients. The two compounds 5-azacytidine and 5-aza-2'-deoxycytidine has been permitted by FDA as a specific treatment for myelodysplastic disorder (Esteller, 2007). Histone deacetylase (HDAC) inhibitors are other potential anticancer agents aimed to treat a tumor epigenetically. The pleiotropic characteristic of these inhibitors expands their outstanding capacity to prompt differentiation, apoptosis, and cell-cycle arrest. Several first phase clinical trials validated the anticancer potential of HDAC inhibitors. For example suberoylanilide hydroxamic corrosive (SAHA) is the foremost drug of this category, has been permitted by the FDA for the treatment of cutaneous T- cell lymphoma. The loss of monoacetylated H4K16 can be switched by another class of medications that hinder sirtuins (subclass of HDACs that deacylate H4K16). Sitruins inhibitors have potential to re-establish the gene expression of epigenetically silenced tumor-suppressor genes. The potential of sirtuin inhibitors establishes their potential use in chemotherapy (Milne and Denu, 2008).

10.1.2 MONOCLONAL ANTIBODY (MAB) THERAPY FOR ASTHMA AND CHRONIC OBSTRUCTIVE PULMONARY DISEASE (COPD)

Eosinophil has been connected with asthma but their role is still uncertain. The goal of pharmacologic actions for aspiratory circumstances is to decrease symptoms, slow decline or recover lung function, and a decrease in incidence and harshness of exacerbations. Inhaled corticosteroids (ICS) manage significant symptoms and exacerbations of asthma and COPD. Control with these mediators is suboptimal, particularly to patients with severe disease. New biologics recognized corticosteroids as adjunctive treatment for eosinophilic inflammation. It has been established that eosinophils have important function in asthma pathology. Nucala (mepolizumab; anti-interleukin (IL)-5) and Cinquair (reslizumab; anti-IL-5), the second and third biologics respectively accepted for the management of asthma. It has been established that eosinophils decline the lung performance in a set of COPD patients (Nixon et al., 2017).

10.1.3 NANOBODIES: ADVANCED LUNG THERAPY

Local pulmonary distribution of biotherapeutics may offer benefits for the control of lung disease. Circulation of therapeutic entity straight to the lung has various perspectives such as the quick onset of action, decreased systemic introduction, and requirement of a lower dose as well as needleless management. Construction of a protein for gasped distribution is puzzling and desires protein with promising biophysical activity, suitable to withstand the force related to formulation, distribution, and inhalation device. Nanobodies are the least functional particles acquired from a naturally occurring heavy chain-only immunoglobulin. They are extremely stable, soluble, and show biophysical properties specifically well suited for pulmonary delivery. A clinical and preclinical study on antibody transportation through pulmonary route was performed. The study reported the benefits of using Nanobodies for inhalation and its distribution in the lung. ALX-0171 Nanobody is an example of clinical advancement for the management of respiratory syncytial viral infections (RSV) (Van Heeke et al., 2017).

10.1.4 POTASSIUM CHANNEL TARGETED CARDIOVASCULAR THERAPY

Potassium channels play crucial role in biological processes involved in various diseases. Modification in such an important protein due to congenital insufficiencies or unwanted side-effects of mutual medication might lead to then dysfunctions. The heart is one of those tissues where potassium channel plays a vital role. Numerous kinds of potassium channel work in a manner to conserve the cardiac action potency. Compounds that alter the function of the potassium channel in heart cells have been recognized as another class of anti-arrhythmic agents. Different cardiomyopathies show undesired symptoms due to medications that directly or indirectly influence the potassium channels. Thus, it is easy to forecast of the inherited predisposition of recognized potassium-linked cardiac channelopathy (Martínez-Mármol et al., 2008).

10.2   DRUG DISCOVERY

10.2.1   *INNOVATION OF DRUG DISCOVERY*

A wide range of therapeutic science characterizes the term drug discovery as a procedure by which novel drug are presented worldwide to treat daring diseases. Drug discovery research is a newer field. Eighteenth-century discoveries characterized a portion of an irreplaceable establishment in the chemical hypothesis. Avogadro's nuclear hypothesis and periodic table compound has been established. Chemical sciences built up a model to organize the elements according to their valences and nuclear weight. There was also a hypothesis of acids and bases. August Kekule conveyed the revolutionary benzene hypothesis for the aromatic organic molecules. This was the birth time of coal- tar derivatives and dyes (Drews, 2000). Dye chemistry evolution had a reflective influence on medical biology. Selective capability of the dyes for biological sample established the presence of chemoreceptors. Chemotherapy deal with the mixture of medications. It intends to expand the life expectancy or to diminish illness symptoms (Drews, 2000). In 1815, Friedrich Wilhelm Adam Serturner isolated morphine from the opium. In 1848, papaverine was isolated yet its antispasmodic action was not well established till 1917 (Drews, 2000). Analytical chemistry validated the isolation and purification of bioactive constituent from medicinal plants and established the medicinal potential in the nineteenth century. As bioactive property of plants ended up uncovered, majority of the pharmaceutical organizations addressed the challenges in standardization of the process of drug purification. In the middle of 1871 and 1918, Oswald Schmiedeberg, Institute of pharmacology at the University of Strasbourg presented several experimental intellectual basics of pharmacology (Drews, 2000). There is a need to create the inventive associations to help interdisciplinary drug research and its advancement.

10.2.2   *DRUG DISCOVERY IN THE TWENTIETH CENTURY*

In the nineteenth century, 33% of all the death in the USA had three basic causes, pneumonia, tuberculosis (TB), and diarrhea. These causes are exceptionally uncommon today, they are treatable. In 1940, the death rate due to these causes was 1 out of 11. In 2000, the figure was down to 1 for every 25 individuals. Pneumonia remains in the top ten list responsible for the massive death from the above mentioned three diseases. It is presently driven by extra complex conditions, cardiovascular disease (CV), and malignancy. Different variables like enhanced sanitation and vaccination undoubtedly performed an important role in the increase of the expected lifespan. Medication for hypertension, contamination, hyperlipidemia, and malignancy additionally subsidized the recognizable improvement in health and expected lifespan. Drug discovery in the pharmaceutical field began soon after miracle drugs, such as penicillin became available. In the meantime, synthetic organic chemistry had built up the methodologies for large-scale preparation of the synthetic drug.

10.2.3 BIOLOGICAL AND BIOSIMILAR DRUGS

Compound originated from living cells/organisms that are recognized as biological medicines. They consist of highly heterogeneous molecular substances difficult to characterize. Some variations might notice among biological medicines due to alterations in the biological system and manufacturing method. Fundamentally, same and clinically comparable medicine to a biological drug is known as biosimilar drug. Reference or originator medicine is an approved biological medicine derives biosimilar active medicine.

Fundamentally, the biological item and reference item are same. Clinically significant dissimilarity is absent between the biological and reference items in relation to potency, purity, and safety of the product. Biosimilar drugs are different from generic medicine. Biosimilar drugs possess only common synthetic structure and are indistinguishable, in terms of molecular structure to their reference drug (Kumar, 2016). Biological products are commonly produced using living cells or organism. They might produce using biotechnology, acquired from natural sources, or created artificially.

PHS Act (Public Health Service Act) considered protein consider as natural item. A few proteins have been approved for medications (section 505 of the Federal Food Drug and Cosmetic Act) and others are authorized as biologics (section 351 of the PHS Act).

According to BPCI (The biologics price competition and innovation act of 2009) act, a native protein is considered biological product and an organic item. Analytical, animal, and clinical examinations are utilized to assess biological item potential of reference items. Studies revealed minor variations in clinical stability, immunogenicity, toxicity, and pharmacokinetics/pharmacodynamics assessment (Kumar, 2016).

10.2.3.1   COST-SAVING POTENTIAL OF BIOSIMILAR DRUGS

A biologic drug (biologics) is a product, manufactured from living organisms or their parts. It includes tissue, cells, blood, and its components, recombinant protein (insulin, mAb), allergen, vaccine, and genes. They have the capability to treat various cancer, rheumatoid arthritis, multiple sclerosis, Crohn's disease, and additional severe diseases. Although biologics improved the treatment for several diseases but are more costly in terms of price per dose. In recent years FDA (U.S. Food and Drug Administration), a federal organization of the United States, took some effort for the approval of low- cost biosimilar drugs. It is expected that the prize of biosimilar drugs might lower to a lesser extent than small-molecule generics (Mulcahy et al., 2014). This is not theoretical; in fact, the perspective associates with current data and previous research in the U.S. market. A study supported by two driving pharmaceutical organization Sandoz, an organization of Novartis Company, and a division of RAND Corporation (RAND Health) anticipated that biosimilar will prompt a $44.2 billion decrease in a coordinate spend on biologic medications from 2014 to 2024, or about 4% of aggregate biologic spent over a similar period, ranging between $13-$66 billion (Mulcahy et al., 2014).

10.2.3.2   BIOLOGICAL AND BIOSIMILAR MEDICINE REACH TO PATIENTS

It has been recognized that in developing countries, specialists are happy with biosimilar drugs. But the budget maintaining persons in control are not concerned to stay on utilization of such expensive medicines. Sometimes, the list of approved medicine is devoid of biosimilar leading to its difficult access. So even getting a right diagnosis the patient won't able to get the medicine. Some barriers have been also considered to use such medicines that meet standards of EMA, WHO, and FDA. It might include high value, government's poor information and awareness, absence of awareness in clinicians and patients, poor/absence of regulation regarding medications, low/lack of political interest, administration complexity, poor diagnosis, screening, and testing. It's nice to see that the regulation of these medications has updated universally at an adequate level. Various nations have already created or are in process to create laws and rules for these drugs. Different countries are building particular rules regarding to the endorsement of these medications. However, these rules are thought to be as per the international quality. For example, Brazil has used WHO guidelines and modified it for their own regulatory guidelines (Kumar, 2016).

10.2.4   DRUG DISCOVERY DESIRES BOTH TIME AND COST

Researchers from diverse background first make the link between their background and particular diseases. For example, the geneticist analyze the genome and identify the correlation between gene and a pathological condition, a microbiologist acknowledges the communication between pathogen and the body, a chemist, and biochemist use to figure out the isolation of a bioactive compound, a pharmacists establishes various methods for the delivery of substance through a pill or injection. The set of disciplines is immense, and needs time and money. Speaking of time, it frequently takes ten to fifteen years to bring a new drug. It means that drugs which appeared early in the 2000s were initially established in 1985 or 1990. Taking consideration of the scientific aspects, clinical trial takes few years to observe the actions of drug. Modeling and marketing also take much of the time for a drug. Speaking of money, according to an article in Forbes, the normal cost to evolve a drug by pharmaceutical company is over $4 billion (Figure 10.1). AstraZeneca disbursed $12 billion dedicated for the research on each official new drug, while GlaxoSmithKline spent over $8 billion.

On the other hand, other corporations like Novartis AG or Amgen Inc. expended below $4 billion. At this order, it appears like a pretty unsustainable. Applicants for a novel drug to manage a disease theoretically include five thousand to ten thousand chemical compounds. On an average two fifty to one thousand, chemical compounds are sufficient but is still high amount. Commonly about ten of these will suit for tests on humans.

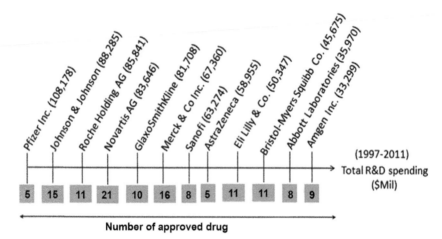

FIGURE 10.1   Representation of cost in million dollars (in brackets) and number of approved drugs by different firms.

Sources: Inno Think Center for Research in Biomedical Innovation; Thomson Research Fundamentals via Fact Set Research Systems.

KEYWORDS

- aromatase inhibitors
- biologics price competition and innovation
- chronic obstructive pulmonary disease
- epidermal growth factor receptor
- inhaled corticosteroids
- monoclonal antibody

REFERENCES

Baskar, R., Lee, K. A., Yeo, R., & Yeoh, K. W., (2012). Cancer and radiation therapy: Current advances and future directions. *Int. J. Med. Sci.*, *9*(3), 193–199.

Drews, J., (2000). Drug discovery: A historical perspective. *Science, 287*(5460), 1960–1964.

Esteller, M., (2007). Cancer epigenomics: DNA methylomes and histone-modification maps. *Nat. Rev. Genet.*, *8*(4), 286–298.

Geyer, C. E., Forster, J., Lindquist, D., Chan, S., Romieu, C. G., Pienkowski, T., Jagiello-Gruszfeld, A., Crown, J., Chan, A., Kaufman, B., & Skarlos, D., (2006). Lapatinib plus capecitabine for HER2-positive advanced breast cancer. *N. Engl. J. Med., 355*(26), 2733–2743.

Johnston, S. R., & Dowsett, M., (2003). Aromatase inhibitors for breast cancer: Lessons from the laboratory. *Nat. Rev. Cancer., 3*(11), 821.

Kumar, S., (2016). Biological/biosimilar drugs: A new hope for better and low cost treatment. *JOJCS, 1*(3), 1–2.

Martínez-Mármol, R., Roura-Ferrer, M., & Felipe, A., (2008). Targeting potassium channels: New advances in cardiovascular therapy. *Recent Pat. Cardiovasc. Drug Discov., 3*(2), 105–118.

Milne, J. C., & Denu, J. M., (2008). The Sirtuin family: Therapeutic targets to treat diseases of aging. *Curr. Opin. Chem. Biol., 12*(1), 11–17.

Mulcahy, A. W., Predmore, Z., & Mattke, S., (2014). The cost savings potential of bio similar drugs in the United States. *Perspectives*, 1–15.

Nixon, J., Newbold, P., Mustelin, T., Anderson, G. P., & Kolbeck, R., (2017). Monoclonal antibody therapy for the treatment of asthma and chronic obstructive pulmonary disease with eosinophilic inflammation. *Pharmacol. Ther., 169*, 57–77.

Sartore-Bianchi, A., Moroni, M., Veronese, S., Carnaghi, C., Bajetta, E., Luppi, G., Sobrero, A., Barone, C., Cascinu, S., Colucci, G., & Cortesi, E., (2007). Epidermal growth factor receptor gene copy number and clinical outcome of metastatic colorectal cancer treated with panitumumab. *J. Clin. Oncol., 25*(22), 3238–3245.

Shiau, A. K., Barstad, D., Loria, P. M., Cheng, L., Kushner, P. J., Agard, D. A., & Greene, G. L., (1998). The structural basis of estrogen receptor/coactivator recognition and the antagonism of this interaction by tamoxifen. *Cell, 95*(7), 927–937.

Van Heeke, G., Allosery, K., De Brabandere, V., De Smedt, T., Detalle, L., & De Fougerolles, A., (2017). Nanobodies® as inhaled biotherapeutics for lung diseases. *Pharmacol. Ther., 169*, 47–56.

Van, C. E., Peeters, M., Siena, S., Humblet, Y., Hendlisz, A., Neyns, B., Canon, J. L., Van, L. J. L., Maurel, J., Richardson, G., & Wolf, M., (2007). Open-label phase III trial of panitumumab plus best supportive care compared with best supportive care alone in patients with chemotherapy-refractory metastatic colorectal cancer. *J. Clin. Oncol., 25*(13), 1658–1664.

Yarden, Y., & Sliwkowski, M. X., (2001). Untangling the ErbB signaling network. *Nat. Rev. Mol. Cell Biol., 2*(2), 127.

CHAPTER 11

Drug Development: *In Silico, In Vivo,* and System Biology Approach

PREM PRAKASH KUSHWAHA,[1] REBATI MALIK,[1]
SHIV GOVIND RAWAT,[2] ATUL KUMAR SINGH,[1] and
SHASHANK KUMAR[1]

[1]*School of Basic and Applied Sciences, Department of Biochemistry, Central University of Punjab, Bathinda, Punjab–151001, India, Tel.: +91 9335647413, E-mail: shashankbiochemau@gmail.com (S. Kumar)*

[2]*Department of Zoology, Banaras Hindu University, Varanasi–221005, Uttar Pradesh, India*

ABSTRACT

Drug development comprises a series of steps to uncover the various properties of a single compound. Various drug development programs have been initiated due to the failure of drugs in clinical trials. Recent approaches encompass various sequential ways to overcome the cost and time consumption in drug discovery. *In silico* approaches includes the screening of compounds, identification of suitable pharmacokinetic activity and their behavior (inhibitory/inducible) against the target. Compounds pass from the *in silico* barriers ready for the testing for various assays and in animal models. Later, successive lead drugs entered in the clinical trials. In this chapter, we reviewed the different software and databases requires for the *in silico* studies and different approaches involved in the drug discovery.

11.1　MODERN MOLECULAR APPROACHES FOR DRUG DISCOVERY: AN INTRODUCTION

Current edge is very exciting for medical science and chemists are involved in all the areas of biomedical research. Chemistry is thus fundamentally well-situated to have a chief impact on drug discovery as other fields could not produce novel molecules. Classical natural product-based drug discovery involves the extraction, functional fraction based assays, isolation, characterization, and target validation. These have been steadily substituted by molecular target-based drug discovery. Computer-based *in silico* high-throughput screening of enormous libraries results in identifying and optimizing hit compounds. The methodologies are conventional since the last two decades (Ojima, 2008). Natural products-based drugs are still the main entities among FDA approved drugs (57.7% of all drugs). Combinatorial chemistry in the form of equivalent synthesis or DOS (diversity-oriented synthesis) for the optimization of highly auspicious lead compounds has been fruitful in numerous drug discovery and developmental cases. The library approach is particularly beneficial to filter the for ADME/T requirement. Structural, chemical, and computational biology, as well as chemical genetics are applied in drug discovery through target-based methodologies. With these advanced tools in hand, using combinatorial chemistry for focused libraries, rational drug designing is now possible. Designing molecular hybrids with a dual-mode of action is a decent example and delivers a hopeful way in the field of modern drug discovery. New antimicrobial drugs in contradiction of multidrug- resistant (MDR) bacterial strains are evolving with the comprehensive usage of modern implements (Ojima, 2008). The terms "biology-oriented synthesis (BIOS)" and "function-oriented synthesis (FOS)" have been arise recently as a reasonable development in chemical genetics. These types of synthetic techniques discover the intrinsic diversity and complications of the structure of natural products. Considerations of benefits of a specific gene with the application of combinatorial biosynthesis offer another charming way in recent drug discovery (Ojima, 2008). Now it has been accepted that organic, medicinal, material, and nanochemistry are involved extremely in the system of drug delivery development.

　　Carbon nanotubes and polymers are used novel as tool for drug delivery. Tumor-specific monoclonal antibodies, omega-3 fatty acids, vitamins, and aptamers have been emerged as extremely promising methods of successful chemotherapy. These strategies have minimum adverse side effects in the targeted drug delivery. Aptamers are oligonucleotide/peptide units combined to a special target molecule typically chosen from a big random sequence pool (Figure 11.1).

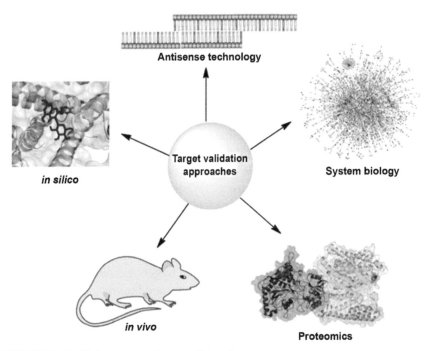

FIGURE 11.1 Various approaches to validate the targets.

For example, synthetic DNAs/RNAs bearing specific affinity for a protein are good aptamers. Molecular designed nano-biomaterial delivers exclusive extracellular matrices individually in different models of human disease (Ojima, 2008). These new materials and disease models help us to execute preclinical efficacy and toxicological study of a specific drug.

11.1.1 IN SILICO TECHNOLOGY

In silico approaches helps us to recognize drug targets with the help of bioinformatics tools (Figure 11.1). This approach has the potential to determine the desired structure for probable binding, generate candidate molecules, specific active site detection, drug-likeness prediction, drug-protein docking, etc. All aspects of the computational method permeate with supercomputers. Workstations form the core of structure-based of drug designing (Wadood et al., 2013). Extraordinary performance computing data handling software and internet give access for the enormous volume of data. It also offers transformation of enormous complex biological information into feasible information and thus aid the knowledge in the present drug discovery procedure. The use of corresponding experimental, as well as bioinformatics methods, enhanced the success probability of various steps involved in drug discovery procedure. *In silico,* studies provide novel targets and particular purposes of the lead compounds with desired pharmacological properties. Computational techniques provide the delivery of new drug candidates quickly at a lower cost. Virtual screening, de novo designing and *in silico* ADME/T modeling and modern methods to study protein-ligand interaction are the primary contributions of computation based drug discovery (Wadood et al., 2013).

11.1.1.1   SIGNIFICANCE OF IN SILICO DRUG DESIGN

In the current scenario, additional protein targets become accessible via nuclear magnetic resonance, crystallography, and bioinformatics methods. There is a serious need of computational tools that can recognize and investigate the active sites on protein and drug molecules having potential to interact with these specific sites. Time and money necessary for the planning of a new drug are massive and intolerable. Some data interpreted that it takes about $880 million cost and 14 years of research time to create a new drug, before it is announced in the market. Interference of computer at some reasonable steps is important to lower the time and cost necessary in the journey of drug discovery (Wadood et al., 2013).

11.1.1.2   ACCESSORIES FOR IN SILICO MODEL INTERACTIONS

Computer simulations are fast and comparatively cheap option for preliminary screening of both active targets and potential drugs. De novo Pharmaceuticals of Cambridge, UK, has made a set of software for the virtual screening, docking as well as for ligand-based designing (Table 11.1). If the active target has a known structure, the site explorer can predict the potential drug binding sites and estimation of the interactions between active sites and drug. For an active target whose structure is unknown, Quasi2 software is applied to create a virtual protein for significant binding experiment (Wadood et al., 2013). The SkelGen suite package can use its own data to create a new chemical structure enhanced for the interaction with a targeted active site. The pharmaceutical companies are cooperating with GeneFormatics in San Diego, California, in an agenda dedicated to find inhibitors of M10 family of matrix metalloproteinases enzymes. These enzymes involved in inflammatory and cancer ailments. GeneFormatics is using proteomics to recognize the target enzymes and illustration of their active sites while De novo is docking/virtual screen model for small molecules against the target protein. Software that can model ligand-receptor communications are can acquired from a number of firms including Accelrys in San Diego, California; Tripos of St Louis, Missouri; and Metaphorics of Mission Viejo, California. A number of software are free access for researchers at non-profit organizations such as GOLD established by Cambridge crystallographic data center (CCDA), UK, and AutoDock 3 by Scripps research institute at La Jolla, California. Molsoft in La Jolla, California, designed ICM molecular modeling software and also released an ICM browser for Apple Macintosh (Wadood et al., 2013). Another suite for structural homology program is, Accelrys. It identifies the potential function, fold family, and 3D structure of target proteins by matching them with the sequences and structural homologs of their known function. After identification of protein structure, the functional information can be collected by different modules within Accelrys's Insight II program which supports various processes, for example, X-ray crystallography, NMR studies, and protein modeling. LION bioscience target engine in Heidelberg, Germany, supports target prediction by subscription. It has the potential to examine the gene sequence and expression data, map potential functional features of protein structure, find homologous structures, view related gene annotation and protein pathway information and uses text mining to find out the functional interactions. Proteins themselves developed as active drugs in biotherapeutics. A software suite designed for the optimization protein function is protein design automation (PDA) produced by Xencor. PDA has the ability to screen huge numbers of amino-acid changes in an acknowledged protein structure. It initiates functional information from 3D protein structure and design novel features into the protein to enhance its function (Wadood et al., 2013).

TABLE 11.1 Acknowledgment of Numerous Tools and Databases Involved *in Silico* Methodology

Databases		
Chemical databases	Zinc database	A group of commercially accessible chemical compounds with three-dimensional coordination delivered from the Shoichet Laboratory of the Department of Pharmaceutical Chemistry situated in UCSF.
	Zinc15 Database	ZINC database is comprised of 100 plus million purchasable compounds in ready to dock, three-dimensional formats delivered from the Shoichet Laboratory of the Department of Pharmaceutical Chemistry situated in UCSF.
	ChEMBL	Database of small molecules Comprises interaction and functional possession of small molecules binding to their macromolecular targets.
	ChemSpider	ChemSpider is assembly of chemical compounds established by Royal Society of Chemistry. It comprises the transformation of chemical name into chemical structure, creation of SMILES (simplified molecular input line entry system), InChI (international chemical identifier) strings, and calculation of numerous physicochemical parameters.
	Drug Bank	Cheminformatics and bioinformatics resource combines the precise drug data with complete drug target information. Permits searching for comparable compounds.
	PubChem	Database of chemical compounds sustained by National Center for Biotech Information (NCBI), along with bioassays results. Permits analogous compounds search (2D and 3D).
	Approved Drugs	It comprises thousands of chemical arrangements and also designations for small molecular drugs permitted according to the US Food and Drug Administration (FDA). Structures or names can be searched in a list of names, sorted by structural features, and ranked by resemblance to user-drawn structure.
	Structural Database (CSD)	It contains crystal structure for small molecules in CIF format. CSD is organized and sustained through Cambridge Crystallographic Data Center (CCDA).

TABLE 11.1 (Continued)

Protein- ligand complexes databases	Protein data bank (PDB)	Databank of experimentally finalized structure of nucleic acids, proteins, and complex assemblies.
	Protein-ligand database (PLD)	Collection of protein-ligand complexes derived from PDB with further biomolecular data, comprises binding energies, Tanimoto ligand equality scores with protein sequence, resemblances of protein-ligand complexes and shared by the University of Cambridge.
Pathway databases	STITCH	To discover known and expected communications among proteins and other chemicals. Chemicals are joined to other proteins and derived from experiments and databases. STITCH covers interactions for three lakh molecules and also 2.6 million proteins discovered in 1133 organisms. Dispersed by the biotech center of TU Dresden.

Chemical Structure Representations

2D drawing	ChemDraw	Molecule editor established by the cheminformatics firm Cambridge Soft.
	MarvinSketch	Innovative chemical reviser for drawing chemical structures, reactions, and queries established by ChemAxon.
2D drawing online	ChemWriter	Chemical structure editor designed to utilize with Web applications. Dispersed by Metamolecular.
	TouchMol Web	Able to make biological and chemical structures online. It permits copy/paste to ISIS/Draw, ChemDraw, and SciFinder as well as Word and has capability to convert name-to-structure. Delivered by Scilligence
3D viewers	UCSF Chimera	Open source, highly extensible package for co-operative analysis and visualization of molecular structures and its related data related to it. Free for government, academician, profitless, and personal uses. Established by the resource for Visualization, Biocomputing, and Informatics, UCSF.

TABLE 11.1 (Continued)

	Pymol	User-sponsored, open source, molecular visualization system transcribed in Python. Dispersed by DeLano Scientific LLC.
	Visual molecular dynamics (VMD)	Free open source molecular visualization program to present, animate, and examine huge biomolecular systems using three-dimensional graphics and in-built scripting. Established by NIH resource for macromolecular modeling and bioinformatics, University of Illinois.
File format Converters	OpenBabel	Permitted open source chemical expert software. Mostly used to convert chemical file formats.
	iBabel	iBabel is another different graphical network to Open Babel.
Analysis of ligand-protein interactions	DS Visualizer	Free three-dimensional visualizer created by Discovery Studio. Permits sequence handling and two or three-dimensional charting. It creates 2-dimensional ligand-receptor interaction figures. Shared by Accelrys.
Web services	E-Babel	Online form of OpenbBabel. Distributed through Virtual Computational Chemistry Laboratory.
	e-LEA3D	Draw a molecule by using the ACD applet (v.1.30) and create three-dimensional coordinates with the assistance of Frog program.
Molecular Modeling		
Software	CHARMM	CHARMM (Chemistry at HARvard macromolecular mechanics) is a package for molecular simulation programs, consist of demos, and source code.
	GROMACS	GROMACS (GROningen MAchine for Chemical Simulations) is an open-source for molecular dynamics simulation package.
	Amber	Assisted Model Building along with Energy Refinement is a kit of molecular simulation program containing demos and source code.
Web Services	SwissParam	Conveys parameters and topology for small-scale organic molecules supported by CHARMM (Chemistry at HARvard macromolecular mechanics). All atoms force field can be used with GROMACS and CHARMM.

TABLE 11.1 *(Continued)*

	R.E.D. Server	Targeted service to automatically derive ESP and RESP charges and to create force field libraries for novel molecules/molecular parts.
		Homology Modeling
Software	Modeler	Software for creating homology representation of tertiary protein confirmations, using a procedure stimulated by nuclear magnetic resonance called satisfaction of spatial restraints. Sustained by Andrej Sali of the UCSF. Free for academics. Commercial versions and Graphical user interfaces are shared by Accelrys.
	I-TASSER	Internet service is valuable for protein function and structure prediction. Models are created depending upon multiple- threading alignments by iterative TASSER and LOMETS simulations. I-TASSER (as 'Zhang-Server') was classified as the topmost server in recent CASP8 and CASP7 experiments. Distributed through the University of Michigan.
Web services and databases	SWISS-MODEL	Fully computerized protein structure homology-modeling server, available through the ExPASy web server, or through the program DeepView (Swiss PDB-Viewer).
	Robetta web server	Rosetta homology modeling and abinitio fragment assemblage with Ginzu domain estimation
	FOBIA	Implemented through the structural Bioinformatics group of Tel-Aviv University
		Binding Site Prediction
Software	Fpocket	Open-source protein pocket (cavity) finding algorithm works on Voronoi tessellation. Presently accessible as a command line determined program and established in C language programming. fpocket comprises two different programs (dpocket and tpocket) that allows us to find pocket descriptors and assess own scoring functions respectively. Also possesses a druggability prediction score.
	GHECOM	Program for conclusion of multi-scale pockets located on protein exterior by utilizing mathematical morphology. Free of cost open source.

TABLE 11.1 *(Continued)*

Databases	Pocketome	Encyclopedia of a structural assemblage of every druggable binding site which can be recognized experimentally through the co-crystal structures in protein data bank.
	PocketAnnotate database	Database formed by the non-redundant binding sites from every existing protein-ligand complexes obtained from PDB. Redundancy was lowered as the best possible binding site for a given ligand is elected (by taking into consideration, the highest resolution structure) per fold of a protein. Sustained by the Department of Biochemistry, IISc.
Web services	DoGSiteScorer	Automated pocket analysis and detection web service utilized for protein evaluation. Predictions with DoGSiteScorer deals with the calculations on automatically predicted pockets. These calculations are utilized for druggability evaluation by the software. Sustained by the Hamburg University.
	ConCavity	Used to Predict ligand binding site according to protein sequence and structure.
	SplitPocket	Binding sites prediction for the non-associated structures.

Molecular Docking

Software	Autodock	Free open source EA centered docking software. Makes proteins and ligand side chains flexible. Sustained through the Molecular Graphics Laboratory of The Scripps Research Institute, la Jolla.
	DOCK	Anchor-and-Grow centered docking program. Free of cost for academic usage. Makes proteins and ligand side chains flexible Sustained through the Soichet group of the UCSF.
	GOLD	GA dependent docking program. Flexible ligand. Partial flexibility to protein. Availed by a partnership among the University of Sheffield and GlaxoSmithKline plc as well as CCDC.
Web services	SwissDock	A web service to conclude the molecular connections that are present in a small molecule and a target protein.
	DockingServer	DockingServer offers a web-based, simple to understand interface that takes cares of all procedures of molecular docking commencing protein and ligand set-up.

TABLE 11.1 *(Continued)*

	1-Click Docking	Free online molecular docking program. The solution can be seen online in 3D by WebGL/JavaScript dependant on molecule viewer module of GLmol. Distributed by Mcule.
	Screening	
Software	PyRX	A virtual Screening program for Computational Drug Design utilized to screen libraries of compounds to target potential drugs. PyRx comprises docking wizard with an easy-to-handle user interface that gives it a valuable importance for Computer- Aided Drug Design. PyRx also contain visualization engine and chemical spreadsheet-like functionality, important for the Rational Drug Design. AutoDock Vina and Autodock 4 are utilized as a docking program. Free of cost and open source.
	CATS	Chemically Advanced Template Search. Topological pharmacophore descriptor designed for scaffold-hopping, focused library profiling, screening selection of compounds. Devised by the Swiss Federal Institute of Technology at Zurich (ETHZ).
Web services	e-LEA3D	Able to search the FDA permitted drugs either by substructure or by keyword. Also used to create a combinatorial library of molecules.
	wwLig-CSRre	Online Tool to develop a bank of small compound with alike compounds as per query.
	Target Prediction	
Software	MolScore-Antivirals	The system used to prioritize and identify antiviral drug candidates. Established by Pharma Informatic of Germany.
	MolScore- Antibiotics	The system used to prioritize and identify antibacterial drug candidates. Established by PharmaInformatic of Germany.
Web services	ElectroShape Poly-pharmacology server	Web service for the assessment of side effects of compounds and polypharmacology profiles according to molecular similarity concept. Established and sustained by Alvaro Cortes at Cabrera

TABLE 11.1 *(Continued)*

Ligand Design

Software	LEGEND	The program is used for automated structure-based drug design and utilizes an atom-centered growing approach. Introduced by IMMD.
	Autogrow	Design ligand design using fragment-based growing and docking, as well as evolutionary techniques. AutoGrow utilizes AutoDock as selection operator. Administered through the McCammon Group at UCSD.
Databases	sc-PDB-Frag	Database of protein-bound fragments to assist the selection of truly bioisosteric scaffolds.
Web services	e-LEA3D	Invent concepts of ligand (scaffold-hopping) with the help of de novo drug design software LEA3D.
	3DLigandSite	An automated method to predict binding site for the ligands. Distributed by the Imperial London College London UK.

Binding Free Energy Estimation

Software	X-score	Used to calculate binding compatibility of the ligand molecules to their target protein. X-Score is availed to the public for free.
	NNScore	Use python script to calculate binding free energies from PDBQT files of ligand and the receptor, by utilizing a neural network approach. Free of cost and open source. Distributed through the McCammon Lab of UCSD.
Web services	PreDDICTA	Predicts DNA-drug interaction strength by Calculating ΔTm (melting temperature) and Affinity of binding.
	PharmaGist	Freely available web server utilized for pharmacophore detection. The download version also has the ability of virtual screening.

QSAR

Software	clogP	Program for computing log Poct/water according to the structure Shared by BioByte.

TABLE 11.1 (Continued)

	ClogP/CMR	Estimates Molar Refractivity and logP shared by Tripos.
	Topomer CoMFA	3D QSAR tool used in automated construction of the models for determining the properties of any particular compound or biological activity. Shared by Tripos.
Databases	MOLE db	Molecular Descriptors Data Base is a non-payable on-line database consists of 1124 molecular descriptors estimated for hundreds to thousands of molecules.
Web services	XScore-LogP	Calculates the octanol/water partition coefficient of a drug, using the X-Score program.
	3-D QSAR	3-D QSAR MODELS DATABASE used in virtual screening, construction of own molecules on the server, or upload of already drawn structure. The module also gives space to choose target protein for biological activity prediction and virtual screening.
ADME Toxicity		
Software	ADMEWORKS Predictor	QSAR centered Virtual (*in silico*) screening scheme used for concurrent evaluation of compounds properties.
	MedChem Designer	A tool that associates molecule character with several free ADMET property predictions of ADMET Predictor. Shared through the Simulations Plus, Inc.
Web services	Free ADME Tools	ADME Calculation Toolbox of the SimCYP application made available by free of charge by SimCYP.
Databases	SAR Genetox Database	Genetic toxicity database intended to be utilized as a source to create predictive modeling training groups. Distributed by Leadscope.
	SAR Carcinogenicity Database	Carcinogenicity database for approved structures. It is utilized as a source for making training sets. Shared by Leadscope.

11.1.2   ANTISENSE TECHNOLOGY

Disruption of particular gene expression to decrease the quantity of the respective protein is another way of target validation (Figure 11.1). This also examines the role of the target in normal physiology. The technique includes antisense technology, RNA interference (RNAi) and gene knockouts. Antisense technology uses short oligonucleotides to target particular mRNAs for destruction. This technology was developed as an apparatus to find out oligonucleotide-based drugs that hinder the gene expression rather than protein function. Antisense technology provides huge success as a high throughput technology for validation. It is highly specific and efficient to stop the expression of potential target protein in different experimental models. GeneTrove Company focused on the untouched pool of possible therapeutic for target RNAs validation and drug development. It offers traditional target validation package that involves optimized antisense inhibitors against target of interest and control oligonucleotides. Antisense technology can be used both at *in silico* and *in vivo* for the target validation (Aboul-Fadl, 2006). Biognostik, a biotechnology company offers a drug target validation kit which could be applied *in vivo* or *in vitro*. It comprises five target specific phosphorothioate antisense inhibitors and two random-sequence oligonucleotides to control nonspecific properties. It also established a sequence design organization called RADAR which controls antisense oligonucleotides based on specificity, minimal nonspecific properties, or protein binding and potential to get inside the cells. Sequitur, a functional-genomics company combines an antisense library with high-throughput DNA microarray assays to test the potential of antisense molecules on gene expression. The company's skill was applied to validate an important therapeutic target for Alzheimer's disease. Sequitur also carried out RNAi based target validation. Custom phosphorothioate antisense oligonucleotides can be acquired from various firms for the research purpose (Aboul-Fadl, 2006).

11.1.3   THE PROTEOMICS APPROACH

The major drawback of genetic level target validation is that several genes produce various protein isoforms having slightly different function. So target validation is best done by tempering the protein activity via post-translational modification, not by alteration in their expression levels (Figeys, 2002). This permits researchers to escape unwanted variations in the expression of other proteins (another drawback of genetic manipulations). Developments of protein microarray, multidimensional liquid-based protein separation, and methods that influence protein expression and protein-protein interactions have their own impact. ProCode helps to study the role of a cell's protein make-up.

A ProCode library (developed by Xencor) is a protein-expression library of a cell or tissue of interest. In this library, every translated protein is labeled with a plasmid, a circular, small part of DNA containing its equivalent complementary DNA (cDNA). The expressed protein is screened for its communication with possible drugs and the tags of cDNA permit its easy identification by other protein which gives a positive reaction. Xerion's XCALIbur has potential for functional validation and identification of likely drug targets simultaneously. Target-specific antibodies are used to identify the proteins and CALI (chromophore-assisted laser inactivation) to switch off-target proteins photochemically by changing their efficient sites. XCALIbur has the capability to validate particular targets for individual diseases or search new possible targets with disease-associated functions. It searches hits from a set of antibodies specifically created against the proteome of a diseased cell. This antibody binds near to proteins functional site and contains dye released by CALI resulting into the inactivation of protein's functional sites. If inactivation has a consequence on the disease, the protein is precipitated with the help of attached antibody and examined by mass spectroscopy and database (Figeys, 2002).

11.1.4   IN VIVO APPROACH FOR TARGET VALIDATION

The extreme challenge for the target approval is finding of precise animal models for human disease. *In vivo* target validation, through gene knockout, is an influential method for the study of drug action (Figure 11.1). This type of target validation depends on the hypothesis of silencing the gene for the individual target has the equal consequence as administering an extremely precise inhibitor of the target protein *in vivo*. Effective use of mouse knockout technology, drug discovery activities leads to breakthrough therapeutics (Kirchmair et al., 2014). Zebrafish have recently being considered as an animal model for some human diseases. The fish are affordable, easier to preserve and faster to rise than mammals. Drug toxicity and potential therapeutic activity can be tested more comfortably. Surprisingly, disease-causing genes in zebrafish are very identical to those of humans. For example, gene involved in angiogenesis, inflammation, and insulin regulation are similar in zebrafish and human. Transparent embryo of zebrafish provides easy large-scale and high throughput genetic drug screening. Zygogen in Atlanta, Georgia, has established a transgenic zebrafish system named as Z- Tag which could be utilized for target validation. The firm also makes multiple visible zebrafish organs by tagging the tissues with fluorescent markers (Kirchmair et al., 2014). Another most widely used *in vivo* model is mice but working with mice can be both time-consuming and expensive. Lexicon Genetics industrialized the mouse knockouts by using gene targeting, trapping, and embryonic stem cell technologies. Conventional knockout and transgenic mice are also obtainable from Deltagen in the Redwood City California and MemorecStoffel in Cologne Germany.

11.1.5   SYSTEM BIOLOGY

System biology methodologies initially describe specific targets within their normal biochemical pathway. Approach needs number of measurement of the cells, tissue, and body fluids at several levels including transcript, protein as well as endogenous metabolite and enzyme product (Butcher et al., 2004). System biology approach compares these profiles with similar profiles of the diseased condition and associates the profile features with the disease in question. The result gives a set of potential targets that might disturb by RNAi or transgenic. Consequent biological changes are monitored by the same systems biology approach. Systems biology is a truthful exciting approach for enlightening the contributory relation between target modulation and properties on the disease. It is also capable to predict the specific toxicity by measuring the normal function of the target (Butcher et al., 2004). Another target validation approach is MDS Proteomics it uses bioinformatics, gene- expression information, and high throughput protein analysis as well as protein pathway biology to recognize potential targets and gain an all-around signature of their cellular role.

KEYWORDS

- biology-oriented synthesis
- chemistry at Harvard macromolecular mechanics
- chromophore-assisted laser inactivation
- complementary DNA
- diversity-oriented synthesis
- international chemical identifier

REFERENCES

Aboul-Fadl, T., (2006). Antisense oligonucleotide technologies in drug discovery. *Expert Opin. Drug Discov., 1*(4), 285–288.

Butcher, E. C., Berg, E. L., & Kunkel, E. J., (2004). Systems biology in drug discovery. *Nat. Biotechnol., 22*(10), 1253–1259.

Figeys, D., (2002). Proteomics approach in drug discovery. *Anal. Chem., 74*(15), 412A–419A. Kirchmair, J., Mannhold, R., Kubinyi, H., & Folkers, G., (2014). *Drug Metabolism Prediction.* John Wiley & Sons.

Ojima, I., (2008). Modern molecular approaches to drug design and discovery. *Acc. Chem. Res., 41*(1), 2–3.

Wadood, A., Ahmed, N., Shah, L., Ahmad, A., Hassan, H., & Shams, S., (2013). *In silico* drug design: An approach which revolutionized the drug discovery process. *Drug Des. Delivery., 1*(1), 3–7.

CHAPTER 12

Drug Development: Drug Delivery Carriers and Clinical Trial

PREM PRAKASH KUSHWAHA, SWAGATA DAS, ATUL KUMAR SINGH, and SHASHANK KUMAR

School of Basic and Applied Sciences, Department of Biochemistry, Central University of Punjab, Bathinda, Punjab–151001, India, Tel.: +91 9335647413, E-mail: shashankbiochemau@gmail.com (S. Kumar)

ABSTRACT

Engrossment of genomic information to introduce drug discovery and development has announced lots of concepts since last three decades. Though the previously data's and reports acknowledged various striking new drugs against aggressive diseases, but their side effects, cost adjustment and availability provided unsatisfactory outcomes. Various drugs also remained unimpressive in the different clinical trials. These trials stages define the drug pharmacology, dose range, drug efficacy, therapeutic effect, and long term effect. Failure of drug delivery to the diseased position also hampered its successiveness in the clinical trials. Later, various drug delivery systems come up with their ability to solve the challenges against the drug delivery in body system. In this chapter, we reviewed the various clinical trial phases and several drug delivery systems with their renounced characteristics.

12.1 CLINICAL TRIALS: AN INTRODUCTION

Preclinical improvements begin before clinical trial. This aims to govern the care and efficiency of the intervention. With reference to, the preclinical studies if therapy is safe and operative, clinical trials are initiated. On the basis of clinical trials, the interventions are decided as positive or negative for human subjects. The clinical trial undergo in a stepwise process known as phases 0, I, II, III, IV, and V (Figure 12.1).

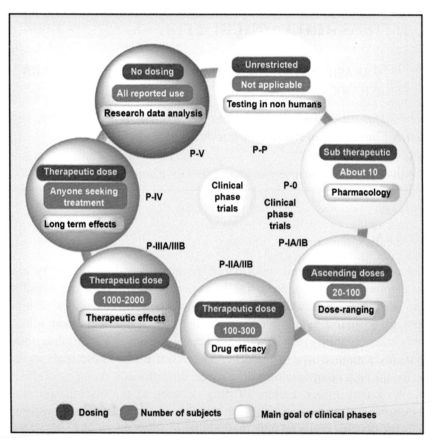

FIGURE 12.1   Clinical phase trials: The stage of drug development.

12.1.1   CLINICAL TRIALS: PHASE 0

The IND (investigational new drug) is also widely known as Phase 0 analyses it provides information on pharmacokinetics as well as selection of promising product from a certain group of candidates followed by evaluation of their bio-distribution and determination of mode of action.

The motive of analysis is to assess the benefit of go versus no-go outcome of early drug. In this early drug, developmental process human models are considered instead of trusting on animal model data. Investigative IND studies shown in early clinical phase includes a restricted human introduction without diagnostic intent. Doses are sub-therapeutic and patients are observed by the clinical investigator and include about 10 study patients for less than week duration. Pharmacodynamics and pharmacokinetics studies are carried out. These trials are based on earlier traditional dose appreciation, safety, and tolerance studies. It neither replace the phase I clinical trials nor specifies whether the therapy has positive influence on the targeted pathology or not. It also helps in the removal of candidate therapy before Phase I studies (Kummar et al., 2007). These trials were established to minimize the difficulties in drug development, discovery of pharmacodynamics and pharmacokinetics profiles of IND, aid in accelerated documentation of auspicious drugs and to lower the input of time and money. The drawbacks of these trials comprise lack of therapeutic commitment, motivation of patients to join, delay, or exclude patients from the other clinical trials having therapeutic commitment, micro-dosing pharmacokinetics, association to a therapeutic dose and obtainability of sensitive analytical methods (Le Tourneau et al., 2009).

12.1.2 CLINICAL TRIAL: PHASE I

A Phase I clinical trial is the finest way to manage occurrence, dose, maximum tolerated dose (MTD), side effects tolerability, pharmaco-kinetics, and pharmacodynamics of a drug. These studies are regulated with the safe treatment trials on 20 to 100 patients checked by the clinical investigator. Doses are amplified if do not showed severe side effects and the patients are tested for the positive outcome of therapy. These gratitude dose studies are availed to regulate the finest and harmless dose that can be administered. It is fraction of the dose that caused damage during animal testing. Unnecessary introduction of subjects to sub-therapeutic doses sustaining safety and rapid accumulation is the main aim of Phase I trials (Storer, 1989). In most of the cases, healthy volunteers (having a certain disease) are required. In normally contract research societies conduct this type of studies and salaries might be given. Testing is performed consecu-tively for every patient or a group of patients with proper data study. Dose toxicity and efficacy curves are determined during the phase and involve single-dose trials (Phase IA), multiple arising dose trials (Phase IB), and food effect studies. These are easy to implement and do not need exceptional software.

12.1.3   CLINICAL TRIAL: PHASE II

A Phase I clinical trial study MTD while Phase II study assesses the possible effectiveness, characterize treatment advantage of the disease. These studies are accomplished on massive groups about 100 to 300 subjects and are intended to evaluate the drug's effectiveness and safety evaluations. The therapeutic doses determined in phase I study administered in the patients and monitored by clinical investigator. Trials are frequently accompanied in a multi-institution setting. Phase II is distributed into Phase IIA (preliminary clinical trials to assess effectiveness and care in designated populations with the disease or disorder to be treated diagnosed or banned) and Phase IIB (the toughest trials designed to validate effectiveness). The Phase II design mainly depends on the excellence and potential of Phase I study. A susceptible feature of both phases includes the type of registered patient. In Phase II trial, elimination of patients involves more criteria than Phase III trial. They are strategized as single and multi-stage clinical trials based upon particular endpoint of the interest. Adaptive clinical trial strategies based on temporarily collected data are used in Phase II clinical trials due to its elasticity and efficacy. This strategy enables the investigator to transform or reshape the trial during the study. Adaptive designs are classified by FDA into well-known and less well-known category (Chow, 2014). Well-understood strategies have been used for years with consistent statistical approach. They are well established and FDA approves these study strategies by reviewing the use of submissions. In less well-understood strategies, relative qualities and boundaries are not completely assessed, devoid of valid established statistical methods. FDA also lacks requisite knowledge about submissions using the study strategies.

12.1.4   CLINICAL TRIALS: PHASE III

Phase III trials include full-scale assessment of treatment aimed to obtain the efficiency of the new treatment in the comparison with its standard treatment. It is an extensive and laborious type of technical clinical study of a new treatment in pre-marketing phase of clinical trial. Additionally, they are the most luxurious and time-consuming trials. The trials might be problematic to design and run. Large groups (100 to 3000 subjects) are recruited and trial include strategies such as randomized skillful trial, uncontrolled trial, historical control, no-randomized concurrent trial, factorial design, and group consecutive designs. Patients are checked by clinical investigator and individual physician. Phase III clinical trials are commonly separated into Phase IIIA and Phase IIIB. Phase IIA trial usually carried out after the establishment of efficiency of the therapy and before the regulatory submission of a new drug application (NDA) or any other dossier. On the other hand, Phase IIIB trials are directed after the submission of NDA or other report but earlier than its endorsement and introduction. Randomized phase III trials are the gold standard indication to endorse new drugs. Drug development-related problems includes limited clinical advantage in large RCT's, estimation of a fruitful trial of Phase III based on Phase II data, drug toxicity determination, strategy of drug combination, and cost of the trial.

12.1.5 CLINICAL TRIALS: PHASE IV

Phase IV trials include all the studies achieved after the drug endorsement and related to approved or accepted indications (Elsäßer et al., 2014). These are the post advertising investigation studies. The focus of the trials is how the drug works in a definite world. Anyone looking for treatment from the physician may be cured with the treatment. The physician displays the results of treatment to the subjects. Effectiveness and discovery of rare or long-term side effects over a relatively higher patient population along with longer duration are assessed, healthcare charges and outcomes are determined, and pharmacogenetics is studied in this phase. New clinical suggestions for a drug may be recognized involving a large number of patients and doctors (Bernabe et al., 2014). The FDA may necessitate a designer to direct a Phase IV trial as a necessity for the drug approval. Few numbers of the study about less than half are finished or even started by designers. It may also lead to the removal of drug from market or controlled to certain signs.

12.1.6   CLINICAL TRIALS: PHASE V

Phase V clinical trials demonstrate relative efficacy and community-based investigation. The investigation is based on the collected data. Patients are not observed but the chief attention is to determine the incorporation of a new therapy into widespread clinical training.

12.2   MARKETING OF A DRUG

The estimated extent of the influence of pharmaceutical industry on medical research and patient care can be easily made through an analysis of its marketing strategies. The strategy is mainly organized in five category based on ascending order of potential harm to consumers: targeted promotion of physicians, direct-to-consumer (DTC) advertising, unethical recruitment physicians, conflicts of interest in research, and clinical trial data manipulation.

12.2.1   PHYSICIANS TARGETED PROMOTIONS

The influence of a physician's prescription pattern is significant as it subconsciously affects the promotion of Drug companies. In 2002, $15.63 billion was expended by pharmaceutical industry on promotional activities such as the free office supply, sales representatives, all-expenses-paid events, and gifts given to physicians (Parker and Pettijohn, 2003). According to Dr. Israel, out of the promotional budget, a sum of about $8,000 to $13,000 was spent on single physician (Israel, 2003). Likewise, a 10 year reported study by interns of seven university hospitals was published in 1990. The study reported that prescription practice was merely changed by the regular contact with sales representative (Israel, 2003). Parker and Pettijohn reported that the interaction between a doctor and pharmaceutical representative affects the possibility of a drug to be included into an approved and insured list and is increased thirteen times than normal situation (Parker and Pettijohn, 2003). Studies showed that promotion of drugs are distinguished by their prescription in lieu of probable fact that an ideal physician provides the most suitable therapy and care to the patient, at best economic value. This results in the falsified theory of patients receiving quality care. Actually, these patients are to afford high treatment costs due to biased prescription of potential drug.

12.2.2 *DIRECT TO CONSUMER ADVERTISING*

Although hefty DTC advertising increased the sale of promoted drugs but it is not a best-suited option in terms of patients in health and cost. The number of patients seeking for medical attention for allergy during 1990 to 1998 rendered about 14 million, raised up to 18 million by 1999. The money corresponds to more than 15% of $1.85 billion was targeted for DTC advertisement of prescribed oral antihistamines. In 1999, sale of top 25 DTC-advertised drugs increased by 43% in comparison to previous year. It was due to their increased prescription raised by 34.2%. This further increased the DTC advertising cost from $2.3 billion (2000) to $7.5 billion (2005). This clearly illustrates that increased expenditures on doing advertisement of a drug give an increased profit due to increased number of prescriptions (Parker and Pettijohn, 2003).

12.3 UNETHICAL RECRUITMENT OF PHYSICIANS

In pharmaceutical industries, physicians are recruited by public relation firms. Endorsement of the clinical study by these physicians is another common problem. Ottawa Citizen, a reputed news source of Canada, reported that Dr. Davis Healy, a psychiatrist, was sent a finished review paper of 12 pages by a company to present it at a forthcoming conference as a sole author. Even though Healy declined the offer, the paper was presented at the conference under the name of another doctor (Spears, 2003). Several other cases were reported in which the physicians are offered to forward papers to medical journals in lieu of a considerable amount of money which is luring for some physicians and are denied by others for ethical, risky, and insecurity issues. Furthermore, with the increase by companies in investment, promotions, the value of these enticing offers could be tempting for physicians to accept and involve in such malpractices. This practice ruins the reputation of medical professionals and also put their reliability at a substantial risk.

12.4   RESEARCHER'S CONFLICTS OF INTEREST

Researchers' financial conflict of interest (COI) has the perspective to influence results in corresponding studies. As mentioned by the ICMJE (International Council of Medical Journal Editors), COI mainly include consultancy, stock ownership, employment, honoraria, and patent licensing, excluding factors like financial relationships depending on grants, fellowships, awards, free equipment or drugs, and authors functioning as member of an advisory board or as a speaker (Friedman and Richter, 2004). Conflict issues can be challenged by writing a letter to Editor of the concerned journal. Interest may be academic, research or financial. Furthermore, COI is also arises when a person has a patent on the item and also publish the same in a journal.

12.5   DATA MANIPULATION

Many reported piece of evidence suggests that various pharmaceutical companies tend to manipulate the clinical data in order to avoid the adverse results. Searle (2000) disclosed that Celebrex (celecoxib) is the first safe drug among COX-2 inhibitors than older NSAIDS (e.g., ibuprofen) for the treatment of gastrointestinal system. The claim was supposedly crucial for both the patient as well as a marketer because approximately 107,000 annual hospitalizations due to gastrointestinal complications were reported for the consumption of arthritis drugs. Due to this, Celebrex became heavily marketed drug and the cost was increased up to rose to $2 per pill (Parker and Pettijohn, 2003). But later in 2004, a Group Health Cooperative of Seattle revised the protocol of the study carried out for FDA authorization. It showed that previously the results of only 6 months have been reported giving affirmative results while the 12-month results did not revealed any difference in gastrointestinal complications from other drugs (Brownlee, 2004; Rennie, 2004). Thus, discriminatory reportage of positive results is another common malpractice carried out by pharmaceutical companies for their benefits and has been reported many times.

12.6   DRUG DELIVERY CARRIERS

Colloidal drug transporter systems like micellar solutions, vesicle, and liquid crystal diffusions along with nanoparticle diffusions containing smaller particles of 10–400 nm diameters tend to show higher potential in drug delivery system. Property of a good carrier desires enhanced drug loading and relief properties, long shelf-life and low toxicity. Microstructure of the system assisted by the incorporated drug may affect its molecular interactions particularly if the drug retains amphiphilic and/or mesogenic assets (Figure 12.2).

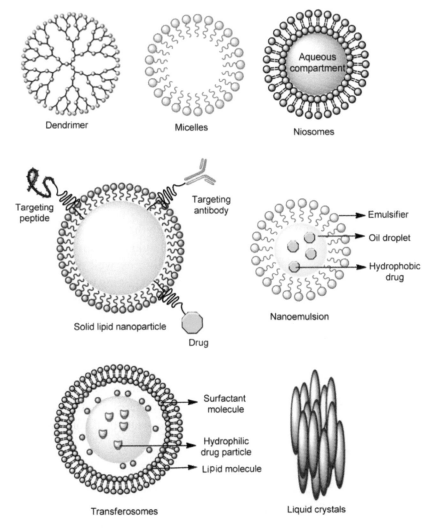

FIGURE 12.2   Different drug delivery carriers.

12.6.1   MICELLES

Micelles designed by the self-assembly of amphiphilic slab copolymers (5–50 nm) in aqueous solution are of great attention for drug delivery claims. These drugs actually can be trapped within the core of slab copolymer micelles and conveyed at concentration that enhances the intrinsic water solubility. Hydrophilic slabs usually create hydrogen bonds with the aqueous environment and tight shell of the micellar core results into formation of hydrophobic core. These properties efficiently protect against hydrolysis as well as degradation by enzymes. A final feature that makes amphiphilic slab copolymers gorgeous for the drug delivery is their easy alternations in chemical alignment, slab length ratios, and total molecular weight permitting permits size and morphology control of the micelles. Functionalization of slab copolymers with cross-linkable groups can upsurge the firmness of the consistent micelles and increase their progressive control.

12.6.2   LIPOSOMES

Liposomes are a system of vesicles that contain either numerous or single phospholipid bilayer. Polar drug molecules are mainly condensed by polar personality of liposomal core. Both amphiphilic, as well as lipophilic molecules, are dissolved in phospholipid bilayer rendering to their affinity near phospholipids. Instead of phospholipids inside the bilayer formation, the contribution of non-ionic surfactants results in niosomes. Drugs that are encapsulated in a nanocage functionalized with channel protein are efficiently threatened from early deprivation by proteolytic enzymes. The drug molecule has diffusion potential through a concentration driven channel which functions on the basis of alterations in interior and exterior concentration of the nanocage.

12.6.3   DENDRIMERS

Dendrimers are nanometer-sized, highly diverged and monodispersed macromolecules having symmetrical construction. They contain an essential core, split unit and terminal functional groups. The core is composed of interior units which regulates the environment of nanocavities and accordingly their solubilizing possessions whereas the exterior groups maintain solubility and chemical performance of these polymers. Targeting efficiency might altered and affected due to interaction of respective ligands at exterior surface of dendrimers while the defense from MPS (mononuclear phagocyte system). Its stability is accomplished by interaction of dendrimers with PEG (polyethylene glycol) chains.

12.6.4 LIQUID CRYSTALS

Liquid crystals are available in both liquid and solid conditions. They can be formed in altered geometries, with alternative polar and non-polar layers (i.e., a lamellar phase) to comprise aqueous drug solutions.

12.6.5 NANOPARTICLES

Nanoparticles (counting nanospheres and nanocapsules of size 10–200 nm) are solid, either shapeless or crystalline. These nanoparticles are capable to encapsulate a drug. This practice protects the drug against chemical and enzymatic deprivation. Nanocapsules are vesicular systems to restrict the drug in a unique polymer membrane-enclosed cavity while nanospheres are generally the matrix system for actual and consistent drug dispersal. Nanoparticles could be composed of both decomposable polymers and non-decomposable polymers. Recently decomposable polymeric nanoparticles concerned attention as possible drug delivery strategies due to their function and application in the controlled announcement of drugs to target particular tissue or organ as the transporters of DNA in gene therapy as well as for their capacity to transport peptides, proteins or genes through the per oral way.

12.6.6 SOLID LIPID NANOPARTICLES (SLN)

Solid lipid nanoparticles (SLN) are particulate matter architecturally connected to polymeric nanoparticles. SLN, unlike the polymeric system, comprises of biocompatible lipids. They are well accepted for physiological *in vivo* administration and do not utilize organic solvents. These lipid matrices contain homolipids like fat or wax that deliver protection to the combined bioactive from chemical and physical deprivation, in addition to alteration in drug relief profile. Typical preparations involve lipids mainly paraffin wax or decomposable glycerides (e.g., Compritol 888 ATO) as the structural base of particle (Attama and Nkemnele, 2005). SLN have an ample and potential application in wide-field including brain, ocular, rectal, oral, topical, and vaccine delivery system, etc. Additionally, it also has better bioavailability, defends sensitive drug molecules from external environment, and even control the release of drug. Common drawbacks of SLN are random gelation tendency, particle growth, unpredicted dynamics of polymorphic transitions and intrinsic low integration rate because of the crystalline structure of solid lipid (Attama and Müller-Goymann, 2008).

12.6.7   NANOSTRUCTURED LIPID CARRIERS (NLC)

Nanostructured lipid carriers (NLC) are colloidal transporters categorized by a solid lipid core containing a solid and liquid mixture of lipids with the range of mean particle size in nanometers. They mainly contain a lipid matrix having a special nanostructure (Nair et al., 2011). This nanostructure recovers drug loading as it decisively recollects the drug during loading. NLC system reduces some difficulties related to SLN such a slow payload and drug exclusion on storage of certain drugs as well as the high content of water in SLN diffusion. Fresher approach to produce NLC is being established.

12.6.8   LIPID DRUG CONJUGATES (LDC) NANOPARTICLES

The main problem with SLN is the low ability to load hydrophilic drugs because of the separating property during construction procedure. Only, extremely effective low dosage of hydrophilic drugs might be appropriately combined inside the solid lipid matrix (Schwarz et al., 1994). To overcome this restriction, LDC nanoparticles with drug loading capabilities of up to 33% were developed. A bulk and insoluble drug-lipid conjugate is primarily prepared mainly by the salt formation or alternatively by covalent linking. The obtained LDC is administered along with an aqueous surfactant solution through high-pressure homogenization or HPH transformed to nanoparticle formulation by HPH. These nanoparticles possess possible application for brain as they might be used as hydrophilic drug in severe protozoan infections (Olbrich et al., 2002).

12.6.9 TRANSFERSOMES

Transfersome machinery was developed with the purpose for providing a vehicle to permit bioactive molecule delivery via the dermal barrier. Transfersomes are basically ultra-deformable liposomes comprising of phospholipids plus additional edge active amphiphiles like bile salt sallow exciting alteration of the vesicle form. The vesicle diameter is about 100 nm when discreted in buffer (Cevc et al., 1998). These flexible vesicles are supposed to permeate through intact dermis layer completely with the help of hydrostatic gradient force existing in the skin. Antigen or drug may be combined into these vesicles in a way parallel to liposomes.

12.6.10 NIOSOMES

Non-ionic bilayer forming surfactants forms niosomes are several vesicles. Niosomes are architecturally similar to liposomes but their synthesized surfactant has an advantage over phospholipids. Niosomes are meaningfully less expensive and have advanced chemical constancy than their naturally occurring phospholipid counterparts (Kazi et al., 2010). Different strategies such as hydration of synthetic non-ionic surfactants or in combination with cholesterol and other lipids are used toniosomes. Niosomes are usually comparable to liposome functionality which mainly deals with the increased bioavailability of drug and as decreased clearance, in a similar manner like liposomes. Niosomes can also be utilized for embattled drug delivery comparable to liposomes. The possessions of niosomes rest both on composition of the bilayer and method of production. Antigens along with minor molecules are transported using niosomes (Lakshmi et al., 2007).

12.6.11 LIQUID CRYSTAL DRUG DELIVERY SYSTEMS

The nanostructured liquid crystalline resources are extremely stable at dilution which makes it easier to continue mainly as a slow and stored source of drug relief in excess body fluids including gastrointestinal tract or subcutaneous space. They may also discrete into nanoparticle form retaining the parent liquid crystalline structure. The rate of drug discharge is connected directly to the nanostructure of the matrix. Lyotropic liquid crystal systems generally contain amphiphilic molecules and solvents which can be categorized into lamellar (Lα), hexagonal, or cubic mesophases. The recent establishment of substantial attention of the lyotropic liquid crystal systems is mainly due to their outstanding vehicles ability for drug delivery (Guo et al., 2010). Out of these, reversed cubic (QII) and hexagonal mesophases (HII) are the most significant and has been widely examined for their capacity to bear the relief of a long range of bioactive compounds including drugs with lower molecular weight, peptide, protein, and nucleic acids.

12.6.12 NANOEMULSIONS

Lipid-based design exists in a large range of elective delivery system like solution, suspension, self- emulsifying system, and nanoemulsions. Oral nanoemulsions offer a very decent substitute because it can recover the bioavailability of hydrophobic drugs by increasing its solubility. They are extensively used for the administration of both BCS class II and class IV drugs. Oral nanoemulsions use safe eatable material for formulation of delivery system. Nanoemulsions have outstanding capability to encapsulate active compounds due to their minor droplet dimension and huge kinetic firmness (Lovelyn and Attama, 2011).

KEYWORDS

- conflict of interest
- direct-to-consumer
- high-pressure homogenization
- lipid drug conjugates
- maximum tolerated dose
- mononuclear phagocyte system

REFERENCES

Attama, A. A., & Müller-Goymann, C. C., (2008). Effect of beeswax modification on the lipid matrix and solid lipid nanoparticle crystallinity. *Colloids Surf. a Physicochem. Eng. Asp.*, *315*(1–3), 189–195.

Attama, A. A., & Nkemnele, M. O., (2005). *In vitro* evaluation of drug release from self micro- emulsifying drug delivery systems using a biodegradable homolipid from *Capra hircus*. *Int. J. Pharm.*, *304*(1/2), 4–10.

Bernabe, R. D., Van, T. G. J., Raaijmakers, J. A., & Van, D. J. J., (2014). The fiduciary obligation of the physician-researcher in phase IV trials. *BMC Med. Ethics.*, *15*(1), 1–8.

Brownlee, S., (2004). Doctors without borders why you can't trust medical journals anymore. *Washington Monthly*, *36*(4), 38–44.

Cevc, G., Gebauer, D., Stieber, J., Schätzlein, A., & Blume, G., (1998). Ultra flexible vesicles, transfersomes, have an extremely low pore penetration resistance and transport therapeutic amounts of insulin across the intact mammalian skin. *Biochim. Biophys. Acta Biomembr.*, *1368*(2), 201–215.

Chow, S. C., (2014). Adaptive clinical trial design. *Annu. Rev. Med.*, *65*, 405–415.

Elsäßer, A., Regnstrom, J., Vetter, T., Koenig, F., Hemmings, R. J., Greco, M., Papaluca-Amati, M., & Posch, M., (2014). Adaptive clinical trial designs for European marketing authorization: A survey of scientific advice letters from the European medicines agency. *Trials*, *15*(1), 1–10.

Friedman, L. S., & Richter, E. D., (2004). Relationship between conflicts of interest and research results. *J. Gen. Intern. Med.*, *19*(1), 51–56.

Guo, C., Wang, J., Cao, F., Lee, R. J., & Zhai, G., (2010). Lyotropic liquid crystal systems in drug delivery. *Drug Discov. Today.*, *15*(23/24), 1032–1040.

Israel, R., (2003). The relationship between physicians and industry: Faustian or symbiotic?: Presidential address. *Am. J. Obstet. Gynecol.*, *188*(6), 1530–1540.

Kazi, K. M., Mandal, A. S., Biswas, N., Guha, A., Chatterjee, S., Behera, M., & Kuotsu, K., (2010). Niosome: A future of targeted drug delivery systems. *J. Adv. Pharm. Technol. Res.*, *1*(4), 374.

Kummar, S., Kinders, R., Rubinstein, L., Parchment, R. E., Murgo, A. J., Collins, J., Pickeral, O., Low, J., Steinberg, S. M., Gutierrez, M., & Yang, S., (2007). Compressing drug development timelines in oncology using phase '0' trials. *Nat. Rev. Cancer.*, *7*(2), 131.

Lakshmi, P. K., Devi, G. S., Bhaskaran, S., & Sacchidanand, S., (2007). Niosomal methotrexate gel in the treatment of localized psoriasis: Phase I and phase II studies. *Indian J. Dermatol. Venereol. Leprol.*, *73*(3), 157.

Le Tourneau, C., Lee, J. J., & Siu, L. L., (2009). Dose escalation methods in phase I cancer clinical trials. *J. Natl. Cancer Inst.*, *101*(10), 708–720.

Lovelyn, C., & Attama, A. A. (2011). Current state of nanoemulsions in drug delivery. *J. Biomater. Nanobiotechnol.*, *2*(05), 626–639.

Nair, R., Arun, K. K. S., Vishnu, P. K., & Sevukarajan, M., (2011). Recent advances in solid lipid nanoparticle based drug delivery systems. *J. Biomed. Sci. Res.*, *3*(2), 368–384.

Olbrich, C., Gessner, A., Kayser, O., & Müller, R. H., (2002). Lipid-drug-conjugate (LDC) nanoparticles as novel carrier system for the hydrophilic antitrypanosomal drug diminazene diaceturate. *J. Drug Target*, *10*(5), 387–396.

Parker, R. S., & Pettijohn, C. E., (2003). Ethical considerations in the use of direct-to-consumer advertising and pharmaceutical promotions: The impact on pharmaceutical sales and physicians. *J. Bus. Ethics*, *48*(3), 279–290.

Rennie, D., (2004). *Drug Manufacturers Facing Pressure to Reveal Study Protocols That Would Show Even Unfavorable Result: Interview with Drummond Rennie and Marc Mora*. NPR News.

Schwarz, C., Mehnert, W., Lucks, J. S., & Müller, R. H., (1994). Solid lipid nanoparticles (SLN) for controlled drug delivery. I. Production, characterization and sterilization. *J. Control. Release, 30*(1), 83–96.

Spears, T., (2003). *Drug Firms Pay Doctors to Sign "Independent" Clinical Studies.* Ottawa Citizen, A1.

Storer, B. E., (1989). Design and analysis of phase I clinical trials. *Biometrics*, pp. 925–937.

CHAPTER 13

Role of Micro RNA in Prostate Cancer Therapy

MOHAMMAD WASEEM,[1] HADIYA HUSAIN,[2] and IMRAN AHMAD[3]

[1]Department of Bioscience and Bioengineering,
Indian Institute of Technology, Kanpur–208016, Uttar Pradesh, India,
Tel.: +91 6386862179, E-mail: mdwaseemsayeed@gmail.com

[2]Biochemical and Clinical Genetics Lab, Section of Genetics,
Department of Zoology, Aligarh Muslim University, Aligarh–202002,
Uttar Pradesh, India

[3]Environmental Toxicology Group, CSIR-Indian Institute of Toxicology
Research, Lucknow–206001, Uttar Pradesh, India

ABSTRACT

MicroRNAs (miRNAs) are small RNAs which are noncoding and regulate gene expression by the process of degradation of target miRNAs at the post-transcriptional level. Deregulation of miRNAs often occurs in cancer and they also govern various other biological activities. The role of miRNAs has substantially intensified as they are prospective biomarkers which have proved to be very useful in the identification and remedial processes of PCa pathogenesis. There are two functionalities of miRNA, oncogenic, and tumor suppression. The changes and modifications that occur during cancer progression in the miRNAs of cancer tissues are related with various medical and etiological factors. In the control and mediation of PCa miRNAs also are likely drug targets and prospective therapeutics. This review targets to scrutinize and explore the developments and breakthroughs attained in the contribution and part of miRNAs in PCa progression.

13.1   INTRODUCTION

The genetic aberrations which cause the tumor suppressor genes to be inactive and activate the oncogenes lead to cancer. In the cancer etiology most of the mutations are acquired hence, they are contemplated to be mechanisms of tumorigenesis. A predisposition to cancer may arise resulting from some of the inherited mutations. The expansion of more developed implements for prognostics, diagnostics, and cure of cancer is the main raison d'être for the ongoing exhaustive researches on the molecular mechanisms of cancers. In this reference, there has been development of novel drugs targeting cancer-related mechanisms in particular. Over the past three decades, the arena of PCa diagnosis has been revolutionized due to the accessibility of the blood test for the prostate-specific antigen (PSA). PSA is discharged into the circulating blood as a repercussion of hampering of the intact prostate architecture. The formation of PSA occurs in healthy prostate gland secretions which are a kallikrein-related serine protease (Lilja et al., 2008). Approximately one-third of the patients diagnosed newly have metastatic PCa despite the fact that the most of cases are localized to prostate. In localized PCa, surgical resection and radiation have been reckoned to be curative. The spreading of cancer from the prostate gland has almost nil treatment solution. Development of novel immunotherapeutic approaches like tumor vaccines has been possible by the discerning of prostate tumor antigens (Miller and Pisa, 2007). The number of cases with advanced PCa is anticipated to elevate as the population ages. For starters the action of androgen deprivation as the initial therapy for patients affected with advanced stage PCa, i.e., metastatic has been acknowledged for over 60 years and the requirement for growth of approximately all metastatic PCa is testosterone (Huggins and Hodges, 1941; Huggins, 1967). Surgical (orchiectomy) or medical castrations help in accomplishment of hormone deprivation. In 16–18 months, all patients advance to a clinical state of androgen independence which results in death while hormonal remedy steers to abrogation which lasts up to 2–3 years (Saitoh et al., 1984; Grayhack et al., 1987; Harada et al., 1992; Bubendorf et al., 2000; Shah et al., 2004; Roudier et al., 2003). Relapsing of a subgroup of patients to grow to a distant metastatic condition may take place while many patients with localized disease require no additional treatment. Patients having metastatic condition are further treated by withdrawing androgenic hormones by the help of orchiectomy or by medical castration using GNRH agonists. Progressive and hormone refractory PCa is the one against which the responses give way while most of the patients respond to hormone ablation (Majumder and Sellers, 2005). There are two main types of PCa according to epidemiological point, sporadic form which is more common than the rare hereditary type. The genes, for example, the ELAC2, MSR1, CHEK2R, NSB1, and NASEL have been discerned as the potential inherited genes susceptible to PCa, however mutations in these genes does not lead up to hereditary PCa in most cases and sporadic cancer cause attributable to these genes are also occasional (Hughes et al., 2005). The primary modulators are the androgens which regulate the prostate in both the normal and cancerous conditions. The androgen-receptor is the one on which the PCa cells are dependent for the primary arbitrator of growth and perseverance in the process of androgen-dependent progression (Feldman and Feldman, 2001; Debes and Tindall, 2004). The enzyme 5a-reductase converts testosterone to dihydrotestosterone; this is a more agile hormone having 5- to 10-fold more affinity towards the receptor of androgen as testosterone enters the cell. Dihydrotestosterone orchestrates the response by attaching to the receptors of androgen in the cytoplasm, which causes phosphorylation, dimerization, and further translocation into the nucleus, subsequently adhering to the androgen- response elements present in the DNA, with ensuing activation of genes that are engrossed in cell survival and development (Feldman and Feldman, 2001). Prostatic acid phosphatases (PAP), PSA, and prostate specific membrane antigen (PSMA) are some of antigens that have already been identified (Harada et al., 2003). Numerous immunotherapy trials which include peptide (Noguchi et al., 2003), DNA (Pavlenko et al., 2004), virus-based (Eder et al., 2004), dendritic cell (Pandha et al., 2004), and genetically-modified tumor cell vaccines (Simons et al., 1999) are the result of the above-identified antigens ventures, which induce tumor-specific T-cells that could be exhibited in some patients (Vieweg and Dannull, 2005). In the aspect of tumor regression and survival, there has been limited success.

13.2 ACTION MECHANISM AND MICRORNA

Small non-coding RNAs are termed as miRNAs that perform the work of regulation of the gene expression at the post-transcriptional level and are approximately 19–22 nucleotides in length (Bartel and Chen, 2004). They are known to be significant moderators of expression of genes (Lewis et al., 2003), which impact apoptosis, differentiation, cell proliferation, and many other physiological functions. Transcription of a longer precursor of miRNA which is called primary-miRNA (pri- miRNA) is the start of the miRNA biogensis (Lee et al., 2004). The Drosha and DGCR8 complex in the nucleus cleave this pri-miRNA to generate a 70-nucleotide stem-and-loop precursor called pre- miRNA (Denli et al., 2004; Gregory et al., 2004; Lee et al., 2003). The pre-miRNA after being transported to the cytoplasm by Exportin-5 gets cut by the RNAse dicer which yields a 20 base pair miRNA/miRNA duplex (Lund et al., 2004). An RNA-induced silencing complex (RISC) is formed by the loading of this duplex onto Argonaute (Ago) proteins (Bernstein et al., 2001). The suppression of the target genes of miRNAs occur either by depleting the constancy of a target mRNA via instigation of its deadenylation and deterioration or by arbitrating translational suppression (Baek et al., 2008; Bagga et al., 2005; Giraldez et al., 2006; Lim et al., 2005; Selbach et al., 2008; Wu et al., 2006).

13.3   ONCOGENIC MICRORNA

Oncogenic miRNAs (oncomiR) are particularly regulated by the tumor suppressor gene and overexpressed in cancer cells. MiR-21 is well known recurrent over-expressed oncomiR in cancer. Many mRNA targets are controlled by it and these targets are related to microvascular proliferation and tumor invasiveness. It has predictive significance for biochemical recurrence risk after radical prostatectomy in PCa patients and the positive expression of miR-21 is congruous with weak biochemical recurrence-free survival (Li et al., 2012). Its expression corresponds with metastatic disease and castration resistance and also augments simultaneously with clinical parameters (lymph node metastasis, Gleason score). Thus, in cancer progression miR-21 is also helpful as a potential biomarker (Cannistraci et al., 2004). These miR-18a, miR-32, miR-106/miR-25, miR-125b, miR-141, miR-221/ miR-222, miR-375, miR-650 well studied miRNA in prostate cancer (PCa).

13.4   TUMOR SUPPRESSOR MICRORNA

Tumor suppressor microRNAs are particularly regulated in the oncogene and these are down expressed in the cancer cell. Several studies have reported the association of miR-34 family as a tumor suppressor in cancer and several approaches have also been devised to replace miR-34 for the remedial process of cancer (Bommer et al., 2007; He et al., 2007). Various signaling molecules which are associated in many stages of PCa advancement may be targeted while miR-34a may take on its tumor suppressor role. miR-34 expression strongly instigated by DNA detrition and oncogenic in a p53-dependent manner. miR-34 activation promote the peculiar attributes of p53 activity (Romano et al., 2016; Chieffi et al., 2009): such as apoptosis through down-regulation of various proteins (CDK4, CDK6, cyclin D1, cyclin E2, E2F3, and BCL2) and instigation of cell cycle block (Chang et al., 2007; Raver-Shapira et al., 2007; Tazawa et al., 2007). PCa tumor progression is suppressed by miR-133 and miR-146a via aiming EGFR, a promoter of tumor for this disease. Therefore, the loss of miR-133 and miR-146a may be ascribed for the amplification of EGFR signaling, which leads to defiant PCa advancement (Sherwood et al., 1998).

13.5 CARCINOGENESIS AND MICRORNA

Emerging evidentiary support points towards the part of miRNAs as either oncogenes or tumor suppressors and the deregulated expression of miRNAs commonly found in cancers of human and tumorigenesis is modulated by the miRNAs through multiple mechanisms. For example, in anti- oncogenic pathways miRNAs have been exhibited to work as oncogenes by inhibition of key molecules. Incitement of oncogenic mechanisms can occur by deletion of a miRNA that aims and suppresses oncogenes (Ventura and Jacks, 2009) various miRNAs have been connected to be as tumor suppressors on the basis of their reduced expression or frequent deletion in numerous human cancers (Croce, 2009). Oncogenic pathways are suppressed by various tumor suppressor miRNAs by the repression of the crucial oncogenes expression. For example, the *RAS* oncogene in lung cancer is reported to be repressed by the members of the *let-7* family (Johnson et al., 2005), reduce the mRNA of *MYC*, leading to decreased proliferation in Burkitt lymphoma cells (Sampson et al., 2007). The chromosomal translocations can occasionally lead to curtailment of the 3' UTRs of oncogenes which lead to the furtherance of oncogenic conversion and loss of miRNA-arbitrated suppression according to recent reports (Mayr et al., 2007).

13.6 DEREGULATION OF MICRORNAS IN PROSTATE CANCER (PCA)

In primary PCa and also CRPC samples there have been frequent reports of the deregulation of miRNAs (Mattie et al., 2006; Porkka et al., 2007; Prueitt et al., 2008). The miRNA expression has been profiled in six cell lines of PCa, four benign prostatic hyperplasia (BPH), ninePCa xenografts samples, and nine PCa tumor samples by Porkka and colleagues. The tumor samples are clustered into hormone-refractory PCa versus hormone naïve PCa by the use of expression profiles of miRNAs and 51 individual miRNAs have been differentially expressed in this study (Porkka et al., 2007). This research not only shows the significance of miRNAs in regulating PCa carcinogenesis, but also advocates that miRNAs can be worked as biomarkers or novel therapeutic targets for identification and categorization of PCa. In PCa carcinogenesis more than 50 miRNAs have been reckoned to be engrossed and the roles of many have been discerned as well, including *miR-15a~16-1, miR-34, miR- 21, miR148a, miR-616, miR-32, miR-200, miR-126, miR-101, miR-221~222miR-330* family and *miR- 125* family.

13.7 MICRORNAS AS MARKERS FOR RESPONSE TO THERAPY

The patients with PCa of localized category who are unable to undergo radical prostatectomy, radiotherapy is the gold standard for them and testosterone suppression, radium-223, radiotherapy, and hormonal remedial procedures are the therapeutics of choice for patients currently dealing with recurrent PCa evolving towards a castration-resistant phenotype (Loblaw et al., 2013). Chemotherapy and immunotherapy are the only options available for men who do not give a respond to first and even secondary hormone therapy, which comprises competitive AR antagonists (enzalutamide) and steroidogenesis inhibitors (abiraterone). The major focus in this arena is disease monitoring, the prognosis of response to therapy and optimization of drug combination which is ensured by the probability of using miRNA quantification for patient-exclusive therapy decisions in case PSA unreliable as a marker for disease progression.

KEYWORDS

- Argonaute
- benign prostatic hyperplasia
- oncogenic miRNA
- primary-miRNA
- prostate-specific antigen
- prostatic acid phosphatases

REFERENCES

Baek, D., Villén, J., Shin, C., Camargo, F. D., Gygi, S. P., & Bartel, D. P., (2008). The impact of micro RNAs on protein output. *Nature, 455*(7209), 64–71.

Bagga, S., Bracht, J., Hunter, S., Massirer, K., Holtz, J., Eachus, R., & Pasquinelli, A. E., (2005). Regulation by let-7 and lin-4 miRNAs results in target mRNA degradation. *Cell, 122*(4), 553–563.

Bartel, D. P., & Chen, C. Z., (2004). Micromanagers of gene expression: The potentially widespread influence of metazoan microRNAs. *Nat. Rev. Genet., 5*(5), 396–400.

Bernstein, E., Caudy, A. A., Hammond, S. M., & Hannon, G. J., (2001). Role for a bidentate ribonuclease in the initiation step of RNA interference. *Nature, 409*(6818), 363–366.

Bommer, G. T., Gerin, I., Feng, Y., Kaczorowski, A. J., Kuick, R., Love, R. E., Zhai, Y., Giordano, T. J., Qin, Z. S., Moore, B. B., & MacDougald, O. A., (2007). p53-mediated activation of miRNA34 candidate tumor-suppressor genes. *Curr. Biol., 17*(15), 1298–1307.

Bubendorf, L., Schöpfer, A., Wagner, U., Sauter, G., Moch, H., Willi, N., Gasser, T. C., & Mihatsch, M. J., (2000). Metastatic patterns of prostate cancer: An autopsy study of 1,589 patients. *Hum. Pathol., 31*(5), 578–583.

Cannistraci, A., Di Pace, A. L., De Maria, R., & Bonci, D., (2014). MicroRNA as new tools for prostate cancer risk assessment and therapeutic intervention: Results from clinical data set and patients' samples. *Bio. Med. Res. Int.*, 1–17.

Chang, T. C., Wentzel, E. A., Kent, O. A., Ramachandran, K., Mullendore, M., Lee, K. H., Feldmann, G., Yamakuchi, M., Ferlito, M., Lowenstein, C. J., & Arking, D. E., (2007). Transactivation of miR-34a by p53 broadly influences gene expression and promotes apoptosis. *Mol. Cell., 26*(5), 745–752.

Chieffi, P., Franco, R., & Portella, G., (2009). Molecular and cell biology of testicular germ cell tumors. *Int. Rev. Cell Mol. Biol., 278*, 277–308.

Croce, C. M., (2009). Causes and consequences of microRNA dysregulation in cancer. *Nat. Rev. Genet., 10*(10), 704–714.

Debes, J. D., & Tindall, D. J., (2004). Mechanisms of androgen-refractory prostate cancer. *N. Engl. J. Med.*, *351*(15), 1488–1490.

Denli, A. M., Tops, B. B., Plasterk, R. H., Ketting, R. F., & Hannon, G. J., (2004). Processing of primary microRNAs by the Microprocessor complex. *Nature*, *432*(7014), 231–235.

Eder, J. P., Kantoff, P. W., Roper, K., Xu, G., Bubley, G. J., Boyden, J., Gritz, L., Mazzara, G., Oh, W. K., Arlen, P., & Tsang, K. Y., (2000). A phase I trial of a recombinant vaccinia virus expressing prostate-specific antigen in advanced prostate cancer. *Clin. Cancer Res.*, *6*(5), 1632–1638.

Feldman, B. J., & Feldman, D., (2001). The development of androgen-independent prostate cancer. *Nat. Rev. Cancer.*, *1*(1), 34–45.

Giraldez, A. J., Mishima, Y., Rihel, J., Grocock, R. J., Van, D. S., Inoue, K., Enright, A. J., & Schier, A. F., (2006). Zebrafish MiR-430 promotes deadenylation and clearance of maternal mRNAs. *Science*, *312*(5770), 75–79.

Grayhack, J. T., Keeler, T. C., & Kozlowski, J. M., (1987). Carcinoma of the prostate. Hormonal therapy. *Cancer*, *60*(S3), 589–601.

Gregory, R. I., Yan, K. P., Amuthan, G., Chendrimada, T., Doratotaj, B., Cooch, N., & Shiekhattar, R., (2004). The microprocessor complex mediates the genesis of microRNAs. *Nature*, *432*(7014), 235–240.

Harada, M., Iida, M. I., Yamaguchi, M., & Shida, K., (1992). Analysis of bone metastasis of prostatic adenocarcinoma in 137 autopsy cases. In: *Prostate Cancer and Bone Metastasis* (pp. 173–182). Springer, Boston, MA.

Harada, M., Noguchi, M., & Itoh, K., (2003). Target molecules in specific immunotherapy against prostate cancer. *Int. J. Clin. Oncol.*, *8*(4), 193–199.

He, L., He, X., Lim, L. P., De Stanchina, E., Xuan, Z., Liang, Y., Xue, W., Zender, L., Magnus, J., Ridzon, D., & Jackson, A. L., (2007). A microRNA component of the p53 tumor suppressor network. *Nature*, *447*(7148), 1130–1134.

Huggins, C. B., & Hodges, C. V., (1941). Studies on prostatic cancer. 1. The effect of castration, of estrogen and of androgen injections on serum phosphatases in metastatic carcinoma of the prostate. *Cancer Res.*, 1, 293–297.

Huggins, C., (1967). Endocrine-induced regression of cancers. *Science*, 156(3778), 1050–1054.

Hughes, C., Murphy, A., Martin, C., Sheils, O., & O'leary, J., (2005). Molecular pathology of prostate cancer. *J. Clin. Pathol.*, *58*(7), 673–684.

Johnson, S. M., Grosshans, H., Shingara, J., Byrom, M., Jarvis, R., Cheng, A., Labourier, E., Reinert, K. L., Brown, D., & Slack, F. J., (2005). RAS is regulated by the let-7 microRNA family. *Cell*, *120*(5), 635–647.

Lee, Y., Ahn, C., Han, J., Choi, H., Kim, J., Yim, J., Lee, J., Provost, P., Rådmark, O., Kim, S., & Kim, V. N., (2003). The nuclear RNase III Drosha initiates microRNA processing. *Nature*, *425*(6956), 415–419.

Lee, Y., Kim, M., Han, J., Yeom, K. H., Lee, S., Baek, S. H., & Kim, V. N., (2004). MicroRNA genes are transcribed by RNA polymerase II. *EMBO J.*, *23*(20), 4051–4060.

Lewis, B. P., Shih, I. H., Jones-Rhoades, M. W., Bartel, D. P., & Burge, C. B., (2003). Prediction of mammalian microRNA targets. *Cell*, *115*(7), 787–798.

Li, T., Li, R. S., Li, Y. H., Zhong, S., Chen, Y. Y., Zhang, C. M., Hu, M. M., & Shen, Z. J., (2012). miR-21 as an independent biochemical recurrence predictor and potential therapeutic target for prostate cancer. *J. Urol.*, *187*(4), 1466–1472.

Lilja, H., Ulmert, D., & Vickers, A. J., (2008). Prostate-specific antigen and prostate cancer: Prediction, detection and monitoring. *Nat. Rev. Cancer, 8*(4), 268–278.

Lim, L. P., Lau, N. C., Garrett-Engele, P., Grimson, A., Schelter, J. M., Castle, J., Bartel, D. P., Linsley, P. S., & Johnson, J. M., (2005). Microarray analysis shows that some microRNAs down regulate large numbers of target mRNAs. *Nature, 433*(7027), 769–773.

Loblaw, D. A., Walker-Dilks, C., Winquist, E., & Hotte, S. J., (2013). Genitourinary cancer disease site group of cancer care Ontario's program in evidence-based care. Systemic therapy in men with metastatic castration-resistant prostate cancer: A systematic review. *Clin. Oncol., 25*(7), 406–430.

Lund, E., Güttinger, S., Calado, A., Dahlberg, J. E., & Kutay, U., (2004). Nuclear export of microRNA precursors. *Science, 303*(5654), 95–98.

Majumder, P. K., & Sellers, W. R., (2005). Akt-regulated pathways in prostate cancer. *Oncogene, 24*(50), 7465–7474.

Mattie, M. D., Benz, C. C., Bowers, J., Sensinger, K., Wong, L., Scott, G. K., Fedele, V., Ginzinger, D., Getts, R., & Haqq, C., (2006). Optimized high-throughput microRNA expression profiling provides novel biomarker assessment of clinical prostate and breast cancer biopsies. *Mol. Cancer, 5*(1), 1–14.

Mayr, C., Hemann, M. T., & Bartel, D. P., (2007). Disrupting the pairing between let-7 and Hmga2 enhances oncogenic transformation. *Science, 315*(5818), 1576–1579.

Miller, A. M., & Pisa, P., (2007). Tumor escapes mechanisms in prostate cancer. *Cancer Immunol. Immunother., 56*(1), 81–87.

Noguchi, M., Kobayashi, K., Suetsugu, N., Tomiyasu, K., Suekane, S., Yamada, A., Itoh, K., & Noda, S., (2003). Induction of cellular and humoral immune responses to tumor cells and peptides in HLA-A24 positive hormone-refractory prostate cancer patients by peptide vaccination. *Prostate, 57*(1), 80–92.

Pandha, H. S., John, R. J., Hutchinson, J., James, N., Whelan, M., Corbishley, C., & Dalgleish, A. G., (2004). Dendritic cell immunotherapy for urological cancers using cryopreserved allogeneic tumor lysate-pulsed cells: A phase I/II study. *BJU Int., 94*(3), 412–418.

Pavlenko, M., Roos, A. K., Lundqvist, A., Palmborg, A., Miller, A. M., Ozenci, V., Bergman, B., Egevad, L., Hellström, M., Kiessling, R., & Masucci, G., (2004). A phase I trial of DNA vaccination with a plasmid expressing prostate-specific antigen in patients with hormone- refractory prostate cancer. *Br. J. Cancer., 91*(4), 688–694.

Porkka, K. P., Pfeiffer, M. J., Waltering, K. K., Vessella, R. L., Tammela, T. L., & Visakorpi, T., (2007). MicroRNA expression profiling in prostate cancer. *Cancer Res., 67*(13), 6130–6135.

Prueitt, R. L., Yi, M., Hudson, R. S., Wallace, T. A., Howe, T. M., Yfantis, H. G., Lee, D. H., Stephens, R. M., Liu, C. G., Calin, G. A., & Croce, C. M., (2008). Expression of microRNAs and protein-coding genes associated with perineural invasion in prostate cancer. *Prostate, 68*(11), 1152–1164.

Raver-Shapira, N., Marciano, E., Meiri, E., Spector, Y., Rosenfeld, N., Moskovits, N., Bentwich, Z., & Oren, M., (2007). Transcriptional activation of miR-34a contributes to p53-mediated apoptosis. *Mol. Cell., 26*(5), 731–743.

Romano, F. J., Rossetti, S., Conteduca, V., Schepisi, G., Cavaliere, C., Di Franco, R., La Mantia, E., Castaldo, L., Nocerino, F., Ametrano, G., & Cappuccio, F., (2016). Role of

DNA repair machinery and p53 in the testicular germ cell cancer: A review. *Oncotarget.*, *7*(51), 85641–85649.

Roudier, M. P., True, L. D., Higano, C. S., Vesselle, H., Ellis, W., Lange, P., & Vessella, R. L., (2003). Phenotypic heterogeneity of end-stage prostate carcinoma metastatic to bone. *Hum. Pathol.*, *34*(7), 646–653.

Saitoh, H., Hida, M., Shimbo, T., Nakamura, K., Yamagata, J., & Satoh, T., (1984). Metastatic patterns of prostatic cancer. Correlation between sites and number of organs involved. *Cancer*, *54*, 3078–3084.

Sampson, V. B., Rong, N. H., Han, J., Yang, Q., Aris, V., Soteropoulos, P., Petrelli, N. J., Dunn, S. P., & Krueger, L. J., (2007). MicroRNA let-7a down-regulates MYC and reverts MYC-induced growth in Burkitt lymphoma cells. *Cancer Res.*, *67*(20), 9762–9770.

Selbach, M., Schwanhäusser, B., Thierfelder, N., Fang, Z., Khanin, R., & Rajewsky, N., (2008). Widespread changes in protein synthesis induced by microRNAs. *Nature*, *455*(7209), 58–63.

Shah, R. B., Mehra, R., Chinnaiyan, A. M., Shen, R., Ghosh, D., Zhou, M., MacVicar, G. R., Varambally, S., Harwood, J., Bismar, T. A., & Kim, R., (2004). Androgen-independent prostate cancer is a heterogeneous group of diseases: Lessons from a rapid autopsy program. *Cancer Res.*, *64*(24), 9209–9216.

Sherwood, E. R., Van, D. J. L., Wood, C. G., Liao, S., Kozlowski, J. M., & Lee, C., (1998). Epidermal growth factor receptor activation in androgen-independent but not androgen-stimulated growth of human prostatic carcinoma cells. *Br. J. Cancer.*, *77*(6), 855–861.

Simons, J. W., Mikhak, B., Chang, J. F., DeMarzo, A. M., Carducci, M. A., Lim, M., Weber, C. E., Baccala, A. A., Goemann, M. A., Clift, S. M., & Ando, D. G., (1999). Induction of immunity to prostate cancer antigens: results of a clinical trial of vaccination with irradiated autologous prostate tumor cells engineered to secrete granulocyte-macrophage colony-stimulating factor using *ex vivo* gene transfer. *Cancer Res.*, *59*(20), 5160–5168.

Tazawa, H., Tsuchiya, N., Izumiya, M., & Nakagama, H., (2007). Tumor-suppressive miR-34a induces senescence-like growth arrest through modulation of the E2F pathway in human colon cancer cells. *Proc. Natl. Acad. Sci. U.S.A.*, *104*(39), 15472–15477.

Ventura, A., & Jacks, T., (2009). Micro RNAs and cancer: Short RNAs go a long way. *Cell*, *136*(4), 586–591.

Vieweg, J., & Dannull, J., (2005). Technology insight: Vaccine therapy for prostate cancer. *Nat. Rev. Urol.*, *2*(1), 44–51.

Wu, L., Fan, J., & Belasco, J. G., (2006). Micro RNAs direct rapid deadenylation of mRNA. *Proc. Natl. Acad. Sci. U.S.A.*, *103*(11), 4034–4039.

CHAPTER 14

Pathogenesis, Molecular Mechanisms of Progression, and Therapeutic Targets of Liver Fibrosis: An Update

HADIYA HUSAIN and RIAZ AHMAD

Biochemical and Clinical Genetics Lab, Section of Genetics, Department of Zoology, Aligarh Muslim University, Aligarh–202002, Uttar Pradesh, India, Tel.: +91-571-2700920/Ext.: 3445, E-mail: ahmadriaz2013@gmail.com (R. Ahmad)

ABSTRACT

Liver fibrosis is the initial feature of any chronic liver injury, which eventually leads to more serious complications like cirrhosis, fulminant liver, hepatocellular carcinoma, etc. This communication peeks into the major elements of both innate and adaptive immune responses to the cause of fibrosis. The association of inflammation and pathogenesis of liver fibrosis has been explored with hepatic stellate cells (HSCs) and other profibrogenic cells in focus. The fibrotic response can perpetuate by changes in inherent properties of structural cells following various processes of proliferation, differentiation, and activation and is ultimately executed by activated extracellular matrix producing myofibroblasts. This review includes the gist of various molecular mechanisms associated with liver fibrosis, including TGF-β/Smad signaling, Wnt pathway, platelet-derived growth factor (PDGF) signaling, chemokine pathways, nicotinamide adenine dinucleotide phosphate (NADPH) oxidase/ oxidant stress and other immune interactions. The developments towards an understanding of mechanisms and its pathways grant various clinically significant openings and opportunities to facilitate drug designing. In this frame of reference, many potential biomarkers and treatment targets are presented by these mechanisms. New breakthroughs and developments in the biological mechanisms of liver fibrosis and therapeutic interventions based on these mechanisms have been thoroughly discussed and highlighted here.

14.1 INTRODUCTION

A diverse array of chronic and acute stimuli for example toxins, medicines, and alcoholic compounds, viral infections, cholestasis, and metabolic diseases cause of a wound healing response known as fibrosis (Anthony et al., 1978; Schppan, 1990). Quantitative and qualitative depositions of extracellular matrix produced by myofibroblasts are characteristic morphological identifications related to liver fibrosis. Healthy liver lacks myofibroblasts, however, an injured liver has an excess of myofibroblasts accumulation which acts as the major effectors of fibro-genesis (Ahmad et al., 2012; Wang et al., 2016). The determination of prognosis can be easily done by early discernment of cirrhosis and meticulate analysis of fibrosis stage as it gives an upper edge for characterization of cases with higher risk of progressive cirrhosis and hepatocellular carcinoma (Motola et al., 2014). Liver fibrosis advances to cause cirrhosis which is a malady best described by altered normal liver architecture and blood flow, development of septae and nodules, portal hypertension, hepatocellular carcinoma culminating into liver failure (Han et al., 2004; Mukherjee and Ahmad, 2015). In humans, there is a major paucity of efficacious and potential therapies which can be translated by the comprehensive understanding of the mechanisms. Specifically, those mechanisms that lead to liver fibrosis by inflammation, hepatocyte injury, activation of myofibroblasts and cholangiocyte proliferation help in the translation of these therapies. However, fibrosis progression can be regressed by keeping in check the causative pathogenic factors (Hauff et al., 2015; Latief and Ahmad, 2017). The prevalence of hepatic fibrosis occurs due to alcohol abuse, chronic viral hepatitis (HBV and HCV), metabolic disorders, obesity, biliary disease, autoimmune hepatitis, incessant contact to toxins and chemicals, parasitic diseases and drug-induced liver diseases (Mormone et al., 2011). There is perpetual advancement in the arena of non-invasive analysis of liver fibrosis over the past decade. Thus, the traditional gold standard in the liver fibrosis diagnostics which is liver biopsy is starting to lose its significant position in clinical practice. Assessment of biochemical properties of liver fibrosis is carried out by serum markers and physical stiffness of the liver by elastography devices, both forms being non-invasive (Chang et al., 2016).

14.2 CAUSES OF LIVER FIBROSIS

14.2.1 CHRONIC VIRAL HEPATITIS

The most common etiologies of liver fibrosis worldwide include chronic hepatitis B and C viruses with an outreach of millions of individuals (Custer et al., 2004). These infections are also the main source of hepatocellular carcinomas. Both the infections have similar pathogenesis of liver damage with the ensuing advancement of liver fibrosis leading to cirrhosis multiple times. When HBV integrated into the host genome, it is possible that lead to alterations in genomic functions or instability of chromosomes however, HCV is unable to integrate into the host genome. Oncogenic properties are exhibited by various HCV proteins which include non-structural proteins, HCV core proteins, and the envelope. There is a decreased threat of hepatocellular carcinoma in HBV infection due to antiviral treatment and vaccination, and also antiviral therapies like ribavirin significantly assist in reducing the risk of hepatocellular carcinoma in HCV (Mormone et al., 2011).

14.2.2 NON-ALCOHOLIC FATTY LIVER DISEASE (NAFLD)

Parallel to the prevalent diseases like obesity and metabolic syndrome epidemic, there is an increased prevalence of NAFLD this is common chronic liver diseases worldwide (Loomba and Sanyal, 2013). NAFLD ranges from simple to severe form of various arrays of symptoms and diseases as non-alcoholic steatohepatitis (NASH) which is the worst form to simple steatosis which characterizes the simpler symptoms. Cirrhosis leading to hepatocellular carcinoma can be potentially developed after NASH (Vernon et al., 2011).

14.2.3 ALCOHOLIC LIVER DISEASE

Alcohol consumption is a major causative factor in the pathogenesis of chronic liver disease that results in alcoholic hepatitis, fatty liver, fibrosis, cirrhosis, and hepatocellular carcinoma (Miller et al., 2011). Alcohol metabolism produces acetaldehyde by the act of alcohol dehydrogenase which further acts to enhance transforming growth factor $\beta 1$ (TGF$\beta 1$)

secretions along with the stimulation of TGF β type II receptor expression in collagen synthesizing hepatic stellate cells (HSCs) in the liver (Anania et al., 1996). The up-regulation of collagen I protein and stimulation of the collagen type 1 alpha 2 chain (COL1A2) promoter is done by both ethanol and acetaldehyde (Anping, 2002). Acetaldehyde causes mRNA upregulation of COL1A1 in cultured human HSC through distinct mechanisms of action in the initial and late responses (Svegliati-Baroni et al., 2005). The signaling pathway mediated through TGFβ1 in the initial time points for the upregulation of COL1A2 expression is different from the one that causes acetaldehyde-induced fibrogenesis (Svegliati-Baroni et al., 2005).

14.2.4 OTHER CAUSES OF LIVER DISEASE

Other causative factors like steatosis and obesity that associated with liver fibrosis which have a very high potential to cause chronic steatohepatitis and nonalcoholic fatty liver disease along with factors like chronic viral hepatitis and alcoholism. Nonalcoholic fatty liver disease has been described even in non-obese people in developing countries (Das et al., 2010). Chronic hepatitis and fibrosis are peculiarly present in metabolic disorders such as hemochromatosis and Wilson's disease (Andersson and Chung, 2007). Mutation in the HFE (high-iron Fe) gene in hereditary hemochromatosis is an underlying cause for the extensive assimilation and amassment of iron in various organs as well as liver (Feder et al., 1996). A mutation in the ATPase (ATP7B) is responsible for the transportation of copper which causes copper build-up in liver that is a characteristic of Wilson's disease or hepatolenticular degeneration (Zhang et al., 2011). Autoimmune hepatitis characterizes aberrant furnishing of human leukocyte antigen class II in hepatocytes and fosters cell-mediated immune responses against the host liver which may also end up in liver fibrosis (Lin et al., 2008). Advanced stage liver fibrosis and portal hypertension have been triggered by parasitic infections for example schistosomiasis (Andersson and Chung, 2007). Obstruction of bile duct causes cholestasis which further causes chronic portal fibrosis finally culminating into cirrhosis. Animal models of experimentally induced hepatic fibrosis by chronic exposure to chemicals such as N- nitrosodimethylamine, carbon tetrachloride (CCl_4) or thioacetamide exhibit typical features of severe hepatic fibrosis (George and Tsutsumi, 2007; Palacios et al., 2008; Domenicali et al., 2009; Husain et al, 2018a).

14.3   ROLE OF VARIOUS CELL TYPES IN LIVER FIBROSIS

14.3.1   HEPATIC STELLATE CELLS (HSCS)

Hepatic fibrosis pathogenesis involves diverse cell types. The residing place of HSCs is the space of Disse which lies between sinusoidal endothelial cells and hepatocytes (Friedman, 2008). Vitamin A storage and pronounced desmin expression are peculiarly related to quiescent HSCs. There are major changes in quiescent HSCs like elevation in the expression of α-smooth muscle actin (α-SMA), loss of vitamin A content thereby gaining myofibroblasts-like structure, enhanced rough endoplasmic reticulum (ER), acquire properties of proliferation, motility, contractility, and pro-fibrogenicity (Gressner, 1996; Husain et al., 2018b). Primary effectors which drive hepatic stellate cell stimulation are considered to be the damage incurred to hepatocytes and Kupffer cell activation (Nieto, 2006; Nieto et al., 2002).

14.3.2   PORTAL FIBROBLASTS

Quiescent portal fibroblasts create a second population of liver cells implicated in portal fibrosis surround the portal connective tissue in healthy liver (Tuchweber et al., 1996). They express markers different from HSCs (e.g., elastin) and are derived from small portal vessels (Li et al., 2007). Proliferation of portal fibroblasts form onion-like configurations around biliary structures acquires a myofibroblast phenotype and accompanies the proliferation of biliary cells. Implication occurs in the early deposition of extracellular matrix (ECM) in portal zones (Desmoulière et al., 2007). There is a general perception that portal fibroblast activation is caused by biliary epithelial cells however, the major players are yet to be identified.

14.3.3   HEPATOCYTES

Apoptosis of hepatocytes commonly occurs during liver injury. Fas partially mediate this process and TNF-related-apoptosis-inducing ligand (TRAIL) may also be involved (Canbay et al., 2004; Mehal and Imaeda, 2010; Sánchez-Valle et al., 2012). There have been reports that activation of kupffer cells and profibrogenic response occurs after the hepatic stellate cell lines engulf the apoptotic bodies of hepatocytes (Canbay et al., 2003a, b). The interaction of hepatocyte DNA with Toll-like receptor 9 (TLR9), expressed in HSCs, mediates HSC activation by hepatocyte-derived apoptotic bodies (Guicciardi and Gores, 2010). Fibrogenic lipid peroxides are also produced by hepatocytes (Novo et al., 2006).

14.3.4   FIBROCYTES

The beginning of fibrocytes is from hematopoietic stem cells and they are capable of differentiation into myofibroblasts. Secretion of growth factors for the promotion of deposition of the extracellular matrix, proliferation of fibrocytes and their itineration to the injured organ occurs in the conditions of tissue damage (Quan et al., 2004; Kisseleva et al., 2006; Strieter et al., 2009). The organ and the type of injury are factors on which the level of differentiation of fibrocytes into myofibroblasts depends as suggested in various studies (Kisseleva et al., 2006; Strieter et al., 2009). Several studies have exhibited that migration of fibrocytes to lymphoid organs is induced by liver injury (Kisseleva et al., 2006), which suggests that extracellular matrix deposition may not be the only result attained by the functioning of these cells.

14.3.5   BONE MARROW-DERIVED MYOFIBROBLASTS

Mesenchymal stem cells are derived from bone marrow may also give rise to a few hepatic myofibroblasts which have the ability for differentiating into lineage-specific cells and are known as multipotent progenitor cells (Kisseleva et al., 2006; Strieter et al., 2009; Yovchev et al., 2009). The circulating mesenchymal stem cells are a population of cells that are distinguishable from hematopoietic-derived fibrocytes, however; it is unclear if there is any contribution of mesenchymal stem cells to extracellular matrix deposition in liver fibrosis (Kisseleva et al., 2006).

14.3.6   EPITHELIAL-MESENCHYMAL TRANSITION (EMT)

The phenotypic transition of fully differentiated epithelial cells to differentiated mesenchymal cells is epithelial-mesenchymal transition.

Epithelial-mesenchymal transition (EMT) may occur in hepatocytes and cholangiocytes and they may acquire mesenchymal features according to liver cell culture studies (Zeisberg et al., 2007; Wynn and Ramalingam, 2012). Epithelial origin of extracellular matrix producing cells is suggested by more recent studies as they provide strong evidentiary support against EMT in the liver to be the source of myofibroblasts (Taura et al., 2010; Wynn and Ramalingam, 2012; Yan et al., 2015).

14.4 AN ACCOUNT OF SIGNALING PATHWAYS IN LIVER FIBROSIS: STRATEGIC INTERVENTIONS

14.4.1 TGF-β/SMAD SIGNALING

Regulation of extracellular matrix metabolism and tissue fibrosis takes place via TGF-β by the overproduction of type I collagen in both mice and humans. In the development of liver fibrosis which is triggered by parasitic infections like *Clonorchis sinensis, Schistosoma, and Echinococcus multilocularis* and other factors as well TGF-β/Smad signaling plays a critical role as reported by some studies (Calabrese et al., 2003; Yan et al., 2015). However, TGF-β signaling can be blocked by SMAD7 by various ways (Calabrese et al., 2003; Lee et al., 2014; Qu et al., 2015), like regulation of Wnt/β-catenin pathway for influencing induced apoptosis, binding TGFβRI for inhibition of the interaction-dependent activation of Smad2 and collaboration with other effectors for induction of TGFβRI degradation (Luo et al., 2014; Wang et al., 2015).

14.4.2 WNT PATHWAY

There are three signaling pathways being classified from the Wnt family of excreted lipid-modified glycoproteins: Necdin-Wnt, noncanonical (β-catenin-independent), and canonical (β-catenin- dependent). Downregulation of peroxisome proliferator-activated receptor γ (PPARγ) is important for the hepatic stellate cell stimulation and differentiation in the Necdin-Wnt pathway. In the development of suppressing adipogenesis, HSCs promote myogenic and neuronal differentiation and express necdin which belongs to melanoma antigen family. The Wnt pathway inhibition mediated by PPARγ for the effective reversal of hepatic stellate cell activation is restored by necdin (Miao et al., 2013; Li et al., 2014; Yu et al., 2016). However, β-catenin-independent planar cell polarity and noncanonical Wnt/Ca^{2+} pathways are ways by which the noncanonical Wnt signaling occurs. Hence, novel insights can be gained into the pathophysiology of liver fibrosis by a comprehensive understanding of Wnt signaling mechanisms.

14.4.3 PLATELET-DERIVED GROWTH FACTOR (PDGF) SIGNALING

In hepatic stellate cell activation PDGF, signaling is well known as characterized pathways (Chang et al., 2016). The receptor subunits of PDGF undergo dimerization followed by phosphorylation of the tyrosine residues in the intracellular domain after the binding of PDGF to its receptors. The mobilization of intracellular calcium ions for the activation of protein kinase C family members results from the activation of Ras-MAPK (Rat sarcoma-mitogen activated protein kinase) pathway signaling via the PI3K-AKT/PKB pathway which is triggered by binding of PDGF to its receptors (Kelly et al., 1991). Cellular proliferation is the cumulative result of all these events. Development of a contractile, fibrogenic phenotype which relates with the degree of fibrosis and inflammation results from rapid induction of β-PDGF receptor is a characteristic of early HSC activation (Failli et al., 1995; Borkham-Kamphorst et al., 2008). The antagonism of PDGF against HSCs is a potential route for an anti-fibrotic strategy as PDGF is the potent mitogen towards the HSCs (Borkham-Kamphorst et al., 2004). In pre-clinical disease models, the pharmacological inhibition of the PDGFR-β chain has proved to be a promising anti-fibrotic strategy (Ogawa et al., 2010; Wang et al., 2010).

14.4.4 CHEMOKINE PATHWAYS

The mobilization of inflammatory responses within various organs is done by chemokines, are a class of small chemotactic molecules with cytokine-like functions (Wasmuth et al., 2010). Several receptors like C-X-C motif chemokine receptor 3 (CXCR3), chemokine receptor type 5 (CCR5), and CCR7 and various chemokines which include C-C motif chemokine ligand 2 (CCL2), CCL3, CCL5, C-X-C motif ligand 1 (CXCL1), CXCL8, CXCL9, and CXCL10 are expressed by HSCs (Sahin et al., 2010). Enhancement of fibrogenesis by elevation in cell number and intensified inflammation are a result of migration of fibrogenic cells to the site of injury which is generally promoted by chemokines. Induction of RANTES (regulated on activation, normal T-cell expressed and secreted) which is a ligand of CCR5 takes place by NFκB (nuclear factor kappa-light-chain-enhancer of activated B) signaling which stimulate the migration and proliferation of HSCs (Schwabe et al., 2003). Significant decrease of liver fibrosis and macrophage infiltration lead to the CCR1- and CCR5-deficient mouse models of fibrosis (CCl4 and bile duct ligation) (Seki et al., 2009).

14.4.5 NADPH OXIDASE (NOX)/OXIDANT STRESS

Lipid peroxidation generates reactive oxygen species (ROS) which further instigates and sustains fibrosis (Macdonald et al., 2001). Cytochrome P450 2E1 is strongly triggered in alcoholics which leads to enhanced levels of ROS and peri-central (zone 3) injury. Liver injury and fibrosis is mediated by NADPH oxidase (NOX) by activation of HSCs, Kupffer cells and macrophages and generation of oxidant stress. NOX also plays a role in angiotensin signaling. As impediment of Ang II synthesis fosters abrogation of liver fibrosis and inflammation, it is also found that Ang II in the liver directly operates on HSCs exhibiting its profibrogenic attribute. This takes place through the activation of NAPDH oxidase complex and production of ROS subsequently (De Minicis and Brenner, 2007) (Figure 14.1).

14.5 CONCLUSIONS

Comprehensive understanding of the mechanisms and pathways which are played an important role in the pathogenesis of the fibrogenic response can help deduce putative and novel therapeutic management for liver diseases where fibrosis acts as a major deteriorating factor. There is an urgent need in the field of liver fibrosis to the identification of potential and effective therapeutic agents for targeting specific pathways in fibrosis is crucial. The quality and life span of patients are increased worldwide, it can be also improved, and impediments of cirrhosis can be abrogated by the prevention and management of hepatic fibrosis. There has been an expansion in our cognizance of the cellular and molecular mechanisms behind liver fibrogenesis as a result of consequential progress in the past many years. Salient targets for drug development like PDGFR, TGF-β, EGF, and VEGF have been exhumed by the determination of activated

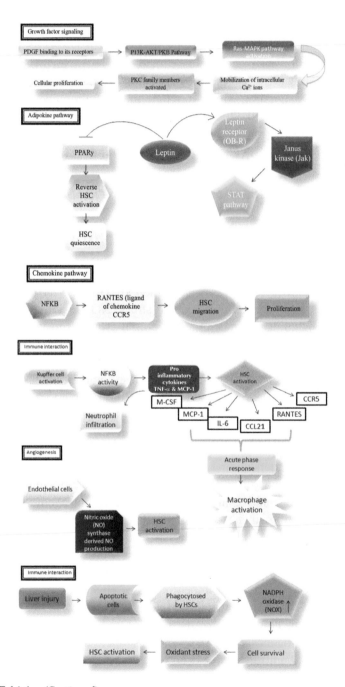

FIGURE 14.1   *(Continued)*

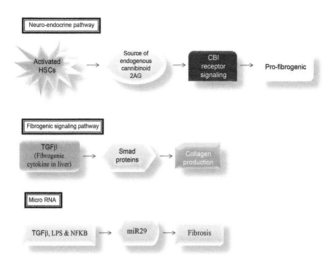

FIGURE 14.1 (a–c): Molecular mechanisms involved in the pathogenesis of liver fibrosis.

HSCs as the key fibrogenic cell type in the injured liver and the divulgence of the facets of hepatic stellate cell instigation and perpetuation. These signals maneuver to generate scar by elevation in the proliferation of HSCs, fibrogenesis, contractility, matrix damage, and pro-inflammatory signaling. The noteworthy new mechanisms of liver fibrosis and prominent signaling pathways that are being actively studied have been explained. For the generation of realistic hopes for effective anti-fibrotic therapies, there have to be unwavering approaches in understanding the exploitation of various pathways for the regression of fibrosis.

ACKNOWLEDGMENT

The authors would like to thank Chairman, Department of Zoology, AMU, Aligarh for providing necessary facilities for this work and to all lab colleagues for timely help and support. Financial assistance to HH (MANF, University Grants Commission, New Delhi, India) is thankfully acknowledged.

CONFLICT OF INTERESTS

The authors declare no conflict of interest.

KEYWORDS

- **hepatocytes**
- **liver fibrosis**
- **myofibroblasts**
- **reactive oxygen species**
- **signaling pathways**
- **vascular endothelial growth factor**

REFERENCES

Ahmad, A., & Ahmad, R., (2012). Understanding the mechanism of hepatic fibrosis and potential therapeutic approaches. *Saudi Journal of Gastroenterology: Official Journal of the Saudi Gastroenterology Association*, *18*(3), 155.

Anania, F. A., Potter, J. J., Rennie-Tankersley, L., & Mezey, E., (1996). Activation by acetaldehyde of the promoter of the mouse α2 (I) collagen gene when transfected into rat activated stellate cells. *Archives of Biochemistry and Biophysics*, *331*(2), 187–193.

Andersson, K. L., & Chung, R. T., (2007). Hepatic schistosomiasis. *Current Treatment Options in Gastroenterology*, *10*(6), 504–512.

Anping, C., (2002). Acetaldehyde stimulates the activation of latent transforming growth factor-β1 and induces expression of the type II receptor of the cytokine in rat cultured hepatic stellate cells. *Biochemical Journal*, *368*(3), 683–693.

Anthony, P. P., Ishak, K. G., Nayak, N. C., Poulsen, H. E., Scheuer, P. J., & Sobin, L. H., (1978). The morphology of cirrhosis. Recommendations on definition, nomenclature, and classification by a working group sponsored by the World Health Organization. *Journal of Clinical Pathology*, *31*(5), 395–414.

Borkham-Kamphorst, E., Kovalenko, E., Van, R. C. R., Gassler, N., Bomble, M., Ostendorf, T., & Weiskirchen, R., (2008). Platelet-derived growth factor isoform expression in carbon tetrachloride-induced chronic liver injury. *Laboratory Investigation*, *88*(10), 1090.

Borkham-Kamphorst, E., Stoll, D., Gressner, A. M., & Weiskirchen, R., (2004). Antisense strategy against PDGF B-chain proves effective in preventing experimental liver fibrogenesis. *Biochemical and Biophysical Research Communications*, *321*(2), 413–423.

Calabrese, F., Valente, M., Giacometti, C., Pettenazzo, E., Benvegnu, L., Alberti, A., Gatta, A., & Pontisso, P., (2003). Parenchymal transforming growth factor beta-1: Its type II receptor and Smad signaling pathway correlate with inflammation and fibrosis in chronic liver disease of viral etiology. *Journal of Gastroenterology and Hepatology*, *18*(11), 1302–1308.

Canbay, A., Feldstein, A. E., Higuchi, H., Werneburg, N., Grambihler, A., Bronk, S. F., & Gores, G. J., (2003). Kupffer cell engulfment of apoptotic bodies stimulates death ligand and cytokine expression. *Hepatology*, *38*(5), 1188–1198.

Canbay, A., Friedman, S., & Gores, G. J., (2004). Apoptosis: The nexus of liver injury and fibrosis. *Hepatology, 39*(2), 273–278.

Canbay, A., Taimr, P., Torok, N., Higuchi, H., Friedman, S., & Gores, G. J., (2003). Apoptotic body engulfment by a human stellate cell line is profibrogenic. *Laboratory Investigation, 83*(5), 655.

Chang, P. E., Goh, G. B. B., Ngu, J. H., Tan, H. K., & Tan, C. K., (2016). Clinical applications, limitations, and future role of transient elastography in the management of liver disease. *World Journal of Gastrointestinal Pharmacology and Therapeutics, 7*(1), 91.

Custer, B., Sullivan, S. D., Hazlet, T. K., Iloeje, U., Veenstra, D. L., & Kowdley, K. V., (2004). Global epidemiology of hepatitis B virus. *Journal of Clinical Gastroenterology, 38*(10), S158–S168.

Das, K., Das, K., Mukherjee, P. S., Ghosh, A., Ghosh, S., Mridha, A. R., Dhibar, T., Bhattacharya, B., Bhattacharya, D., Manna, B., & Dhali, G. K., (2010). Nonobese population in a developing country has a high prevalence of nonalcoholic fatty liver and significant liver disease. *Hepatology, 51*(5), 1593–1602.

De Minicis, S., & Brenner, D. A., (2007). NOX in liver fibrosis. *Archives of Biochemistry and Biophysics, 462*(2), 266–272.

Desmoulière, A., Darby, I., Costa, A. M., Raccurt, M., Tuchweber, B., Sommer, P., & Gabbiani, G., (1997). Extracellular matrix deposition, lysyl oxidase expression, and myofibroblastic differentiation during the initial stages of cholestatic fibrosis in the rat. *Laboratory Investigation; a Journal of Technical Methods and Pathology, 76*(6), 765–778.

Domenicali, M., Caraceni, P., Giannone, F., Baldassarre, M., Lucchetti, G., Quarta, C., Patti, C., Catani, L., Nanni, C., Lemoli, R. M., & Bernardi, M., (2009). A novel model of CCl4-induced cirrhosis with ascites in the mouse. *Journal of Hepatology, 51*(6), 991–999.

Failli, P., Ruocco, C., De Franco, R., et al., (1995). The mitogenic effect of platelet-derived growth factor in human hepatic stellate cells requires calcium influx. *American Journal of Physiology-Cell Physiology, 269*(5), C1133–C1139.

Feder, J. N., Gnirke, A., Thomas, W., Tsuchihashi, Z., Ruddy, D. A., & Basava, A., (2003). The discovery of the new haemochromatosis gene. *Journal of Hepatology, 38*(6), 704.

Friedman, S. L., (2008). Mechanisms of hepatic fibrogenesis. *Gastroenterology, 134*(6), 1655–1669.

George, J., & Tsutsumi, M., (2007). siRNA-mediated knockdown of connective tissue growth factor prevents N-nitrosodimethylamine-induced hepatic fibrosis in rats. *Gene Therapy, 14*(10), 790.

Gressner, A. M., (1996). Trans differentiation of hepatic stellate cells (Ito cells) to myofibroblasts: A key event in hepatic fibro genesis. *Kidney International Supplement, 54*.

Guicciardi, M. E., & Gores, G. J., (2010). Apoptosis as a mechanism for liver disease progression. *In Seminars in Liver Disease, 30*(4), 402.

Han, Y. P., Zhou, L., Wang, J., Xiong, S., Garner, W. L., French, S. W., & Tsukamoto, H., (2004). Essential role of matrix metalloproteinases in interleukin-1-induced myofibroblastic activation of hepatic stellate cell in collagen. *Journal of Biological Chemistry, 279*(6), 4820–4828.

Hauff, P., Gottwald, U., & Ocker, M., (2015). Early to Phase II drugs currently under investigation for the treatment of liver fibrosis. *Expert Opinion on Investigational Drugs, 24*(3), 309–327.

Husain, H., Ahmad, R., Khan, A., & Asiri, A. M., (2018). Proteomic-genomic adjustments and their confluence for elucidation of pathways and networks during liver fibrosis. *International Journal of Biological Macromolecules*.

Husain, H., Latief, U., & Ahmad, R., (2018). Pomegranate action in curbing the incidence of liver injury triggered by Diethylnitrosamine by declining oxidative stress via Nrf2 and NFκB regulation. *Scientific Reports, 8*(1), (8606).

Kelly, J. D., Haldeman, B. A., Grant, F. J., Murray, M. J., Seifert, R. A., Bowen-Pope, D. F., & Kazlauskas, A., (1991). Platelet-derived growth factor (PDGF) stimulates PDGF receptor subunit dimerization and intersubunit trans-phosphorylation. *Journal of Biological Chemistry, 266*(14), 8987–8992.

Kisseleva, T., Uchinami, H., Feirt, N., Quintana-Bustamante, O., Segovia, J. C., Schwabe, R. F., & Brenner, D. A., (2006). Bone marrow-derived fibrocytes participate in pathogenesis of liver fibrosis. *Journal of Hepatology, 45*(3), 429–438.

Latief, U., & Ahmad, R., (2017). Herbal remedies for liver fibrosis: A review on the mode of action of fifty herbs. *Journal of Traditional and Complementary Medicine*.

Lee, J. H., Jang, E. J., Seo, H. L., Ku, S. K., Lee, J. R., Shin, S. S., Park, S. D., Kim, S. C., & Kim, Y.W., (2014). Sauchinone attenuates liver fibrosis and hepatic stellate cell activation through TGF-β/Smad signaling pathway. *Chemico-Biological Interactions, 224*, 58–67.

Li, W., Zhu, C., Li, Y., Wu, Q., & Gao, R., (2014). Mest attenuates CCl4-induced liver fibrosis in rats by inhibiting the Wnt/β-catenin signaling pathway. *Gut and Liver, 8*(3), 282.

Li, Z., Dranoff, J. A., Chan, E. P., Uemura, M., Sévigny, J., & Wells, R. G., (2007). Transforming growth factor-β and substrate stiffness regulate portal fibroblast activation in culture. *Hepatology, 46*(4), 1246–1256.

Lim, Y. S., Oh, H. B., Choi, S. E., Kwon, O. J., Heo, Y. S., Lee, H. C., & Suh, D. J., (2008). Susceptibility to type 1 autoimmune hepatitis is associated with shared amino acid sequences at positions 70–74 of the HLA-DRB1 molecule. *Journal of Hepatology, 48*(1), 133–139.

Loomba, R., & Sanyal, A. J., (2013). The global NAFLD epidemic. *Nature Reviews Gastroenterology and Hepatology, 10*(11), 686.

Luo, L., Li, N., Lv, N., & Huang, D., (2014). SMAD7: A timer of tumor progression targeting TGF-β signaling. *Tumor Biology, 35*(9), 8379–8385.

Macdonald, G. A., Bridle, K. R., Ward, P. J., Walker, N. I., Houglum, K., George, D. K., & Ramm, G. A., (2001). Lipid peroxidation in hepatic steatosis in humans is associated with hepatic fibrosis and occurs predominately in acinar zone 3. *Journal of Gastroenterology and Hepatology, 16*(6), 599–606.

Mehal, W., & Imaeda, A., (2010). Cell death and fibrogenesis. *In Seminars in Liver Disease, 30*(3), 226.

Miao, C. G., Yang, Y. Y., He, X., Huang, C., Huang, Y., Zhang, L., & Li, J., (2013). Wnt signaling in liver fibrosis: Progress, challenges and potential directions. *Biochimie., 95*(12), 2326–2335.

Miller, A. M., Horiguchi, N., Jeong, W. I., Radaeva, S., & Gao, B., (2011). Molecular mechanisms of alcoholic liver disease: Innate immunity and cytokines. *Alcoholism: Clinical and Experimental Research, 35*(5), 787–793.

Mormone, E., George, J., & Nieto, N., (2011). Molecular pathogenesis of hepatic fibrosis and current therapeutic approaches. *Chemico-Biological Interactions, 193*(3), 225–231.

Motola, D. L., Caravan, P., Chung, R. T., & Fuchs, B. C., (2014). Noninvasive biomarkers of liver fibrosis: Clinical applications and future directions. *Current Pathobiology Reports*, *2*(4), 245–256.

Mukherjee, D., & Ahmad, R., (2015). Dose-dependent effect of N′-Nitrosodiethylamine on hepatic architecture, RBC rheology and polypeptide repertoire in Wistar rats. *Interdisciplinary Toxicology*, *8*(1), 1–7.

Nieto, N., (2006). Oxidative-stress and IL-6 mediate the fibrogenic effects of rodent Kupffer cells on stellate cells. *Hepatology*, *44*(6), 1487–1501.

Nieto, N., Friedman, S. L., & Cederbaum, A. I., (2002). Cytochrome P450 2E1-derived reative oxygen species mediate paracrine stimulation of collagen I protein synthesis by hepatic stellate cells. *Journal of Biological Chemistry*.

Novo, E., Marra, F., Zamara, E., Di Bonzo, L. V., Caligiuri, A., Cannito, S., Antonaci, C., Colombatto, S., Pinzani, M., & Parola, M., (2006). Dose dependent and divergent effects of superoxide anion on cell death, proliferation, and migration of activated human hepatic stellate cells. *Gut.*, *55*(1), 90–97.

Ogawa, S., Ochi, T., Shimada, H., Inagaki, K., Fujita, I., Nii, A., & Masferrer, J. L., (2010). Anti-PDGF-B monoclonal antibody reduces liver fibrosis development. *Hepatology Research*, *40*(11), 1128–1141.

Palacios, R. S., Roderfeld, M., Hemmann, S., Rath, T., Atanasova, S., Tschuschner, A., Gressner, O. A., Weiskirchen, R., Graf, J., & Roeb, E., (2008). Activation of hepatic stellate cells is associated with cytokine expression in thioacetamide-induced hepatic fibrosis in mice. *Laboratory Investigation*, *88*(11), (1192).

Qu, Y., Zong, L., Xu, M., Dong, Y., & Lu, L., (2015). Effects of 18α-glycyrrhizin on TGF-β1/Smad signaling pathway in rats with carbon tetrachloride-induced liver fibrosis. *International Journal of Clinical and Experimental Pathology*, *8*(2), (1292).

Quan, T. E., Cowper, S., Wu, S. P., Bockenstedt, L. K., & Bucala, R., (2004). Circulating fibrocytes: Collagen-secreting cells of the peripheral blood. *The International Journal of Biochemistry and Cell Biology*, *36*(4), 598–606.

Sahin, H., Trautwein, C., & Wasmuth, H. E., (2010). Functional role of chemokines in liver disease models. *Nature Reviews Gastroenterology and Hepatology*, *7*(12), 682.

Sánchez-Valle, V., Chavez-Tapia, N. C., Uribe, M., & Méndez-Sánchez, N., (2012). Role of oxidative stress and molecular changes in liver fibrosis: A review. *Current Medicinal Chemistry*, *19*(28), 4850–4860.

Schppan, D., (1990). Structure of the extracellular matrix in normal and fibrotic: Collagen and glycoprotein. *Semin. Liver Dis.*, *10*, 1–14.

Schwabe, R. F., Bataller, R., & Brenner, D. A., (2003). Human hepatic stellate cells express CCR5 and RANTES to induce proliferation and migration. *American Journal of Physiology- Gastrointestinal and Liver Physiology*, *285*(5), G949–G958.

Seki, E., De Minicis, S., Gwak, G. Y., Kluwe, J., Inokuchi, S., Bursill, C. A., & Schwabe, R. F., (2009). CCR1 and CCR5 promote hepatic fibrosis in mice. *The Journal of Clinical Investigation*, *119*(7), 1858–1870.

Strieter, R. M., Keeley, E. C., Hughes, M. A., Burdick, M. D., & Mehrad, B., (2009). The role of circulating mesenchymal progenitor cells (fibrocytes) in the pathogenesis of pulmonary fibrosis. *Journal of Leukocyte Biology*, *86*(5), 1111–1118.

Svegliati-Baroni, G., Inagaki, Y., Rincon-Sanchez, A. R., Else, C., Saccomanno, S., Benedetti, A., Ramirez, F., & Rojkind, M., (2005). Early response of α2 (I) collagen to acetaldehyde in human hepatic stellate cells is TGF-β independent. *Hepatology, 42*(2), 343–352.

Taura, K., Miura, K., Iwaisako, K., Österreicher, C. H., Kodama, Y., Penz-Österreicher, M., & Brenner, D. A., (2010). Hepatocytes do not undergo epithelial-mesenchymal transition in liver fibrosis in mice. *Hepatology, 51*(3), 1027–1036.

Tuchweber, B., Desmouliere, A., Bochaton-Piallat, M. L., Rubbia-Brandt, L., & Gabbiani, G., (1996). Proliferation and phenotypic modulation of portal fibroblasts in the early stages of cholestatic fibrosis in the rat. *Laboratory Investigation; a Journal of Technical Methods and Pathology, 74*(1), 265–278.

Vernon, G., Baranova, A., & Younossi, Z. M., (2011). Systematic review: The epidemiology and natural history of non-alcoholic fatty liver disease and non-alcoholic steatohepatitis in adults. *Alimentary Pharmacology and Therapeutics, 34*(3), 274–285.

Wang, P., Koyama, Y., Liu, X., Xu, J., Ma, H. Y., Liang, S., Kim, I. H., Brenner, D. A., & Kisseleva, T., (2016). Promising therapy candidates for liver fibrosis. *Frontiers in Physiology, 7*, 47.

Wang, Y., Gao, J., Zhang, D., Zhang, J., Ma, J., & Jiang, H., (2010). New insights into the antifibrotic effects of sorafenib on hepatic stellate cells and liver fibrosis. *Journal of Hepatology, 53*(1), 132–144.

Wang, Z. H., Zhang, Q. S., Duan, Y. L., Zhang, J. L., Li, G. F., & Zheng, D. L., (2015). TGF-β induced miR-132 enhances the activation of TGF-β signaling through inhibiting SMAD7 expression in glioma cells. *Biochemical and Biophysical Research Communications, 463*(3), 187–192.

Wasmuth, H. E., Tacke, F., & Trautwein, C., (2010). Chemokines in liver inflammation and fibrosis. *In Seminars in Liver Disease, 30*(3), 215–225.

Wynn, T. A., & Ramalingam, T. R., (2012). Mechanisms of fibrosis: Therapeutic translation for fibrotic disease. *Nature Medicine, 18*(7), 1028.

Yan, C., Wang, L., Li, B., Zhang, B. B., Zhang, B., Wang, Y. H., Li, X. Y., Chen, J. X., Tang, R. X., & Zheng, K. Y., (2015). The expression dynamics of transforming growth factor-β/Smad signaling in the liver fibrosis experimentally caused by Clonorchis sinensis. *Parasites and Vectors, 8*(1), 70.

Yovchev, M. I., Zhang, J., Neufeld, D. S., Grozdanov, P. N., & Dabeva, M. D., (2009). Thymus cell antigen-1-expressing cells in the oval cell compartment. *Hepatology, 50*(2), 601–611.

Yu, F., Lu, Z., Huang, K., Wang, X., Xu, Z., Chen, B., & Zheng, J., (2016). MicroRNA-17–5p- activated Wnt/β-catenin pathway contributes to the progression of liver fibrosis. *Oncotarget, 7*(1), 81.

Zeisberg, M., Yang, C., Martino, M., Duncan, M. B., Rieder, F., Tanjore, H., & Kalluri, R., (2007). Fibroblasts derive from hepatocytes in liver fibrosis via epithelial to mesenchymal transition. *Journal of Biological Chemistry, 282*(32), 23337–23347.

Zhang, S., Chen, S., Li, W., Guo, X., Zhao, P., Xu, J., Chen, Y., Pan, Q., Liu, X., Zychlinski, D., & Lu, H., (2011). Rescue of ATP7B function in hepatocyte-like cells from Wilson's disease induced pluripotent stem cells using gene therapy or the chaperone drug curcumin. *Human Molecular Genetics, 20*(16), 3176–3187.

CHAPTER 15

Prostate Cancer: Molecular Aspects, Biomarkers, and Chemo-Preventive Agents

RISHA GANGULY,[1] ABHAY K. PANDEY,[1]
PREM PRAKASH KUSHWAHA,[2] and SHASHANK KUMAR[2]

[1]Department of Biochemistry, University of Allahabad,
Allahabad–211001, India

[2]School of Basic and Applied Sciences, Department of Biochemistry,
Central University of Punjab, Bathinda, Punjab–151001, India,
Tel.: +91 9335647413, E-mail: shashankbiochemau@gmail.com

ABSTRACT

Prostate cancer (PCa) causes a large number of deaths in men. Age is a pertinent risk factor with increasing probability after the age of 40 years. Other risk factors include high consumption of saturated fatty acids as well as a lack of minerals and vitamins in diet. Around 5% PCas cases are inherited either from father or brother. Prostate-specific antigen (PSA) and acid phosphatase (AP) are the first biomarkers used in the detection of PCa. Besides these, other biomarkers used commonly in prognosis of PCa include AMACR, glutathione-S-transferase (GSTP1), chromogranin A, sarcosine, serum calcium, etc. Inflammation in the prostatic epithelial cell and DNA damage mediated by oxidative stress play a significant role in progression of PCa. Dietary elements present in fruits and vegetables like phenols, flavonoids, indoles, and minerals like zinc and selenium help in the chemoprevention of PCa. In the present chapter, the pathology, diagnosis with the help of biomarkers and advanced therapeutic approaches in PCa has been elaborated. The influence of dietary elements in the prevention and cure of PCa is also discussed.

15.1   INTRODUCTION

The prostate gland is a part of male reproductive system which helps in the production and storage of seminal fluid. The prostate gland increases in size as the men advance in age. Due to mutation or other reasons, cell division exceeds cell death in the prostate gland consequently resulting in uncontrolled tumor growth known as prostate cancer (PCa). PCa results in highest number of deaths in men due to cancer, second only to lung cancer (Parkin et al., 2005). Due to increase in life expectancy, the population of elderly men in India was 5% in 2003 and is estimated to mount to 15% by the year 2050 which might lead to tremendous increase in the occurrence of PCa in men (Imran et al., 2011).

15.2   PATHOLOGY, DIAGNOSIS, AND THERAPY

The incidence of PCa in developed countries has considerably increased in the last few years. Age is a crucial factor in the implication of PCa. There is an exponential rise in the progression of PCa with increasing age after 40 years (Young et al., 1981). Other factors include high consumption of saturated fats, and lack of selenium, vitamin D and vitamin E in the diet (Brawley and Parnes, 2000). Reports have shown that around 5% of all PCa cases are hereditary, coming either from the father or brother (Hayes et al., 1995). PCa arises after a series of genetic mutational events like loss of tumor- suppressive genes such as p53 which is mutated in up to 64% of tumors and p21 in up to 55% (Burton et al., 2000). The recently identified p73 tumor suppressor gene which is significantly homologous to p53 also appears to be mutated in PCa. However, the most common mutation in PCa occurs in tumor suppressor gene MMAC1/p10 which leads to the attainment of metastatic phenotype (Teng et al., 1997). In addition, the over expression of bcl-2 proteins and mutant p53 gene, and augmentation of androgen receptors also lead to impaired hormonal phenotype (Apakama et al., 1996). Only 4% of all PCas result from transitional epithelium of urinary ducts which is called transitional cell carcinoma while more than 95% cases are adenocarinomas. The PCa cell cytoplasm has elevated amounts of prostate-specific antigen (PSA) and AP which act as biomarkers and help in the identification of prostatic tumor cells and differentiate them from other cancerous cells. Early PCa is often asymptomatic as the tumor arises usually from the peripheral region of the prostate. Some of the symptoms of prostatism may include weight loss, haematuria, perineal volume, cachexia, bone pain, and neurological complications. Transrectal biopsy of prostate tissue is commonly used in the detection of PCa (Engelstein et al., 994). Magnetic resonance imaging (MRI) and computed tomography of the pelvis region are also used to detect the extent of local spread of PCa. Some of the widely used therapeutic measures in the treatment of PCa include orchidectomy, radiation therapy, and use of estrogen derivatives, antiandreogens, and gonadotrophin-releasing hormone (GnRH) analogs, as part of hormone therapy. However, reports have shown that radiation or hormone therapy alone is not completely curative. Recent clinical trials have demonstrated that combination of radiation and hormone therapy is more effective in enhancing the survival rate for patients with locally advanced PCa (Shroeder, 1999).

15.3 ROLE OF INFLAMMATION AND DNA DAMAGE IN PROSTATE CANCER (PCA)

Inflammation has been implicated as a carcinogenic insult in a number of human cancers. It leads to reactive species and free radical generation which cause oxidative damage in DNA resulting in mutations (Marzo et al., 2007). Therefore, inflammation of prostate gland is a major risk factor in development of PCa. NKX3.1, a prostate-specific tumor suppressor gene, is a link between mechanisms of normal prostatic cell differentiation and abnormal cellular proliferation during PCa. Studies have reported that NKX3.1 maintains the normal state of prostatic differentiation, while its loss signifies an initial state of prostate carcinogenesis (Bhatia-Gour et al., 1999). Recently, in the *in vitro* studies, it has been recognized that the expression of the NKX3.1 protein is minimal in regions of inflammatory degeneration and in pre-invasive stages of PCa. They also suggest that inflammatory cytokines accelerate loss of NKX3.1 protein function due to rapid ubiquitination along with proteasomal degradation (Mark et al., 2008). The ataxia telangiectasia mutated (ATM) kinase protein acts as a transducer of the DNA damage signal. ATM has several direct targets, including 53BP1 and BRCA1. The ATM-dependent response to DNA damage can block progression through cell cycle before, during, and after DNA replication (Zhi et al., 2010). ATM is also activated due to oxidative stress, independent of DNA damage. Oxidative stress is majorly responsible for genetic damage in the prostate gland, hence ATM plays a significant role in the protection of prostatic epithelial cells from oxidative stress-mediated DNA damage (Ross et al., 2013) (Figure 15.1).

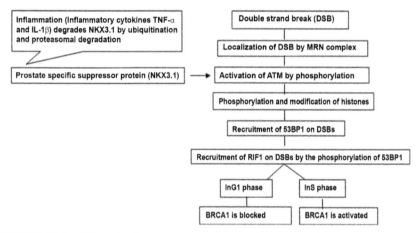

FIGURE 15.1   Effect of inflammatory cytokines on the DNA repair pathway.

15.4   PROSTATE CANCER (PCA) BIOMARKERS

Biomarkers for the detection of PCa include DNA, RNA, and protein-based biomarkers (Figure 15.2). Prostatic acid phosphatase (PAP) also known as serum acid phosphatase (AP) was reportedly the first biomarker found in blood serum for PCa. This was followed by the discovery of PSA produced from the prostate gland as a biological marker for PCa. Some of the most important biomarkers of PCa are discussed in the chapter.

15.4.1   α-METHYLACYL COENZYME A RACEMASE (AMACR)

AMACR is a peroxisomal enzyme which is involved in β−oxidation of branched-chain fatty acids and is known to function as a growth promoter in PCa, independent of androgens (Kuefer et al., 2002; Zha et al., 2003). The AMACR gene is over-expressed in PCa tissue at the levels of messenger RNA and protein and is a highly specific tissue biomarker used in the pathological diagnosis (Jiang et al., 2002, 2004; Rubin et al., 2004).

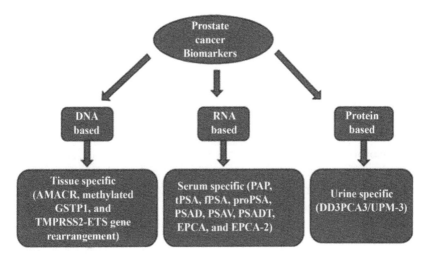

FIGURE 15.2   Prostate cancer biomarkers.

15.4.2   GLUTATHIONE S-TRANSFERASE P1 (GSTP1)

GSTs are a family of multifunctional enzymes which conjugate reduced glutathione (GSH) with reactive substrates and aid in detoxification (Hayes and Pulford, 1995). GSTP1 is a specific PCa tissue biomarker. The GSTP1 gene is reported to be unmethylated in normal tissues but hypermethylated at the promoter region in PCa tissues (Lee et al., 1994; Harden et al., 2003).

15.4.3   CHROMOGRANIN A

Chromogranin A belongs to the granin family of proteins. It is formed in large amounts in neuroendocrine cell types and is encoded by the CHGA gene in humans (Simpson and Aunis, 1989; Berruti et al., 2005). The growth of PCa is regulated by peptides which are derived from. It has been reported that GRN-A possess clinical potential to be a diagnostic tool for PCa (Deftos, 1998).

15.4.4   SARCOSINE

Sarcosine is an N-methyl derivative of glycine. It is a naturally occurring amino acid in muscular and other tissues. Sreekumar et al. (2009) reported that sarcosine stimulates cancerous growth of benign PCa cells and may be used as an indicator of malignancy if detected in urine (Sreekumar et al., 2009).

15.4.5   CAVEOLIN-1 (CAV-1)

Caveolin-1 (Cav-1) is an integral membrane protein which is expressed in two isoforms-caveolin-1α and caveolin-1β. It plays an essential role in regulation of a number of signaling pathways and regulating intracellular processes, particularly as a negative regulator in numerous mitogenic pathways and in oncogenesis (Okamoto et al., 1998; Tirado et al., 2010). Cav-1 which is secreted by PCa cells participates in the process of carcinogenesis. It has been revealed that Cav-1 is over- expressed in PCa cells and is coupled with the gradual progression of the disease (Tahir et al., 2008).

15.4.6   SERUM CALCIUM

Studies have revealed an association between high serum calcium levels and the potential risk of PCa which occurs due to increase in proliferation of cells and inhibition in apoptosis of PCa cells (Lallet- Daher et al., 2009). Researchers postulated that elevated serum calcium is associated to the possibility of terminal PCa. This association was confirmed by another independent group of researchers and they proposed that serum calcium can be a potential prospective biomarker for PCa (Liao et al., 2006; Schwartz et al., 2009).

15.4.7   PCA3/DD3

The PCa antigen 3 (PCA3 or DD3) genes encode a prostate-specific mRNA that is over expressed in cancer prostate tissue (Bussemakers et al., 1999; Kok et al., 2002). Quantitative measurement of PCA3 RNA copies in urine samples containing prostate cells can help in the determination of PCa stages (Haese et al., 2008; Kirby et al., 2009).

15.4.8 KI-67

Ki-67 is a cell-proliferation associated marker and is the only protein which is expressed at levels of cell cycle regulation. It is a very reliable and sensitive biomarker of PCa (Gerdes et al., 1983; Scholzen and Gerdes, 2000).

15.4.9 DAB2IP

DAB2 interacting protein (DAB2IP), a Ras GTPase-activating protein acts as a tumor suppressor. The human DAB2IP gene is located on chromosome 9q33.1-q33.3 and is frequently observed to be downregulated in PCa cell lines (Wang et al., 2002; Chen et al., 2003). It has been reported that down regulation of DAB2IP occurs due to epigenetic alterations like methylation of DNA and histone protein modifications (Chen et al., 2003). The methylation of DNA in the promoter region of DAB2IP gene is known to cause transcriptional silencing and as a result plays a significant role in the gradual progression of PCa. Thus, DAB2IP protein can be used as an effective biomarker for PCa diagnosis (Duggan et al., 2007).

15.4.10 GENETIC BIOMARKERS OF PROSTATE CANCER (PCA)

Genomic analysis is an important tool for evaluation of disease biomarkers. About 70 PCa susceptible loci which account for familial risk (approximately 30%) were identified by the International Practical Consortium (Eeles et al., 2013). Genetic markers such as mutations in BRCA1/2 are well known to be associated with aggressiveness of prostate tumor (Edwards et al., 2010). PCA3, a non-coding RNA is a biomarker expressed only in the prostate is also known as genetic marker. It is detected in urine and the prostatic fluid. In prostate tumor, TMPRSS2:ERG is a very common gene fusion (up to 90%). This gene combination has greater specificity and positive predictive value for PCa. Data from clinical studies suggest that the combination of PCA3 over-expression, TMPRSS2:ERG analysis and serum PSA testing may improve screening effectiveness (Salami et al., 2011). Some of the gene viz., CXCR3, CTAM, KIAA1143, FCRL3, TMEM204, KLF12, and SAMSN1 are known to distinguish between PCa patients and healthy individuals (Velonas et al., 2013).

15.5   DIETARY SOURCE IN CHEMOPREVENTION OF PROSTATE CANCER (PCA)

Several researches have demonstrated the role of fruits, vegetables, and plant-based dietary supplements including vitamins and minerals in the prevention or in reduced incidence of PCa (Sporn and Suh, 2002). Some of the dietary phytochemicals, which are present in fruits and vegetables, have been reported to possess potential anticarcinogenic effects. These include several agents like vitamins, carotenoids, dietary fiber, glucosinolates, isothiocyanates, selenium, indoles, phenols, flavonoids, plant sterols, and protease inhibitors (Vainio, 1999; Greenwald et al., 2002; Vayalil et al., 2004). Diets that include significant amounts of vitamin C rich vegetables and fruits are associated with a low incidence of many forms of cancer (Weisburger, 1999). In animal studies, vitamin C is known to inhibit prostate tumor growth and viability in athymic nude mice transplanted with both androgen- sensitive and -insensitive human PCa cells (Augus et al., 1999; Taper et al., 2001). Studies have shown that combinations of vitamin C and E inhibit the expression of surviving protein, a promoter of PCa cell growth (Gunawardena et al., 2004). Minerals that act in the chemoprevention of PCa are selenium and zinc. Epidemiological studies, preclinical investigations, and clinical trials have revealed that selenium compounds act as potent chemopreventive agents for PCa (Sinha and Bayoumy, 2004). The National Prevention of Cancer Study did a randomized trial on selenium supplementation in 974 men at a dose of 200-µg/day of selenium in 0.5 g high-selenium yeast demonstrated 63% reduction in the incidence of PCa (Clark et al., 1998). Zinc is present in large concentrations in the prostate and is known to inhibit PCa cell growth *in vitro* via inhibition of cell cycle and induction of apoptosis through impaired mitochondrial function (Liang et al., 1999). The plant flavonoids like catechins, apigenin, silymarin, soy isoflavones, and proanthocynadins help in chemoprevention of PCa (Li and Sarkar, 2002).

15.6   CONCLUSION

PCa is a major challenge in males after 40 years of age. Recent studies suggest that inflammatory cytokines and DNA damage are involved in prostate carcinogenesis. The molecular mechanism of inflammation and its possible role in development of PCa are largely ill-defined and unknown. The present chapter gives an idea about the effect of inflammation in prostate cancer. The chapter also evaluates the role of prostate-specific tumor suppressor gene on DNA repair pathway genes. Prevention and cure of prostate cancer using phytochemicals are well known. A variety of prostate cancer preventive agents is derived from plants, marine origin, and/or microbial agents. The medicinal efficacy of many of these agents are well established.

ACKNOWLEDGMENT

R. Ganguly acknowledges financial support from the University Grants Commission, New Delhi for CSIR-UGC Junior Research Fellowship. The authors also acknowledge DST-FIST and UGC-SAP assisted Department of Biochemistry, the University of Allahabad for providing infrastructure facilities.

KEYWORDS

- biomarkers
- dietary supplements
- DNA repair pathway
- glutathione-S-transferase
- inflammation
- prostate cancer

REFERENCES

Agus, D. B., Vera, J. C., & Golde, D. W., (1999). Stromal cell oxidation: A mechanism by which tumors obtain vitamin C. *Cancer Res., 59*, 4555–4558.
Apakama, I., Robinson, M. C., Walter, N. M., Charlton, R. G., Royds, J. A., Fuller, C. E., Neal, D. E., & Hamdy, F. C., (1996). BCl-2 overexpression combined with p53 protein accumulation correlates with hormone-refractory prostate cancer. *Br. J. Cancer, 74*, 1285–1292.

Berruti, A., Mosca, A., Tucci, M., Terrone, C., Torta, M., Tarabuzzi, R., Russo, L., Cracco, C., Bollito, E., Scarpa, R. M., & Angeli, A., (2005). Independent prognostic role of circulating chromogranin A in prostate cancer patients with hormone-refractory disease. *Endocr. Relat. Cancer, 12,* 109–117.

Bhatia-Gaur, R., Donjacour, A. A., Sciavolino, P. J., Kim, M., Desai, N., Norton, C. R., Gridley, T., Cardiff, R. D., Cunha, G. R., & Abate-Shen, C., (1999). Roles for Nkx3.1 in prostate development and cancer. *Genes Dev., 13,* 966–977.

Brawley, O. W., & Parnes, H., (2000). Prostate cancer prevention trials in the USA. *Eur. J. Cancer, 36,* 1312–1315.

Burton, J. L., Oakley, N., & Anderson, J. B., (2000). Recent advances in the histopathology and molecular biology of prostate cancer. *Br. J. Urol. Int., 85,* 87–94.

Bussemakers, M. J., Van, B. A., Verhaegh, G. W., Smit, F. P., Karthaus, H. F., Schalken, J. A., Debruyne, F. M., Ru, N., & Isaacs, W. B., (1999). DD3: A new prostate-specific gene, highly over expressed in prostate cancer. *Cancer Res., 59,* 5975–5991.

Chapman, J. R., Barral, P., Vannier, J. B., Borel, V., Steger, M., Tomas-Loba, A., Sartori, A. A., Adams, I. R., Batista, F. D., & Boulton, S. J., (2013). RIF1 is essential for 53BP1-dependent nonhomologous end joining and suppression of DNA double-strand break resection. *Mol. Cell, 49*(5), 858–871.

Chen, H., Pong, R. C., Wang, Z., & Hsieh, J. T., (2002). Differential regulation of the human gene DAB2IP in normal and malignant prostatic epithelia: Cloning and characterization. *Genomics, 79,* 573–581.

Chen, H., Toyooka, S., Gazdar, A. F., & Hsieh, J. T., (2003). Epigenetic regulation of a novel tumor suppressor gene (hDAB2IP) in prostate cancer cell lines. *J. Biol. Chem., 278,* 3121–3130.

Chen, H., Tu, S. W., & Hsieh, J. T., (2005). Down-regulation of human DAB2IP gene expression mediated by polycomb Ezh2 complex and histone deacetylase in prostate cancer. *J. Biol. Chem., 280,* 22437–22444.

Clark, L. C., Dalkin, B., Krongrad, A., Combs, J. G., Turnbull, B. W., Slate, E. H., Witherington, R., Herlong, J. H., Janosko, E., Carpenter, D., & Borosso, C., (1998). Decreased incidence of prostate cancer with selenium supplementation: Results of a double-blind cancer prevention trial. *Br. J. Urol., 81,* 730–734.

De Kok, J. B., Verhaegh, G. W., Roelofs, R. W., Hessels, D., Kiemeney, L. A., Aalders, T. W., Swinkels, D. W., & Schalken, J. A., (2002). DD3 (PCA 3), a very sensitive and specific marker to detect prostate tumors. *Cancer Res., 62,* 2695–2698.

De Marzo, A. M., Platz, E. A., Sutcliffe, S., Xu, J., Gronberg, H., Drake, C. G., Nakai, Y., Isaacs, W. B., & Nelson, W. G., (2007). Inflammation in prostate carcinogenesis. *Nat. Rev. Cancer, 7,* 256–269.

Deftos, L. J., & Granin, A., (1998). parathyroid hormone-related protein, and calcitonin gene products in neuroendocrine prostate cancer. *Prostate, 8,* 23–31.

Duggan, D., Zheng, S. L., Knowlton, M., Benitez, D., Dimitrov, L., Wiklund, F., Robbins, C., Isaacs, S. D., Cheng, Y., Li, G., & Sun, J., (2007). Two genome-wide association studies of aggressive prostate cancer implicate putative prostate tumor suppressor gene DAB2IP. *J. Natl. Cancer Inst., 99,* 1836–1844.

Edwards, S. M., Evans, D. G., Hope, Q., Norman, A. R., Barbachano, Y., Bullock, S., Kote-Jarai, Z., Meitz, J., Falconer, A., Osin, P., & Fisher, C., (2010). Prostate cancer in

BRCA2 germline mutation carriers is associated with poorer prognosis. *Br. J. Cancer,* *103*, 918–924.

Eeles, R. A., Olama, A. A., Benlloch, S., Saunders, E. J., Leongamornlert, D. A., Tymrakiewicz, M., Ghoussaini, M., Luccarini, C., Dennis, J., Jugurnauth-Little, S., & Dadaev, T., (2013). Identification of 23 new prostate cancer susceptibility loci using the iCOGS custom genotyping array. *Nat. Genetics, 45*, 385–391.

Engelstein, D., Mukamel, E., Cytron, S., Konichezky, M., Sutzki, S., & Vadio, C. S., (1994). A comparison between digitally-guided fine needle aspiration and ultrasound-guided transperineal core needle biopsy of the prostate for the detection of prostate cancer. *Br. J. Urol., 74*, 210–213.

Gerdes, J., Schwab, U., Lemke, H., & Stein, H., (1983). Production of a mouse monoclonal antibody reactive with a human nuclear antigen associated with cell proliferation. *Int. J. Cancer, 31*, 13–20.

Greenwald, P., Milner, J. A., Anderson, D. E., & McDonald, S. S., (2002). Micronutrients in cancer chemoprevention. *Cancer Metastasis Rev., 21*, 217–230.

Gunawardena, K., Campbell, L. D., & Meikle, A. W., (2004). Combination therapy with vitamins C plus E inhibits surviving and human prostate cancer cell growth. *Prostate, 59*, 319–327.

Haese, A., De La Taille, A., Van, P. H., Marberger, M., Stenzl, A., Mulders, P. F., Huland, H., Abbou, C. C., Remzi, M., Tinzl, M., & Feyerabend, S., (2008). Clinical utility of the PCA3 urine assay in European men scheduled, for repeat biopsy. *Eur. Urol., 54*, 1081–1088.

Harden, S. V., Guo, Z., Epstein, J. I., & Sidransky, D., (2003). Quantitative GSTP1 methylation clearly distinguishes benign prostatic tissue and limited prostate adenocarcinoma. *J. Urol., 169*, 1138–1142.

Hayes, J. D., & Pulford, D. J., (1995). The glutathione S-transferase supergene family: Regulation of GST and the contribution of the isoenzymes to cancer chemoprotection and drug resistance. *Crit. Rev. Biochem. Mol. Biol., 30*, 445–600.

Hayes, R. B., Liff, J. M., Pottern, L. M., Greenberg, R. S., Schoenberg, J. B., Schwartz, A. G., Swanson, G. M., Silverman, D. T., Brown, L. M., Hoover, R. N., & Fralmeni, Jr. J. F., (1995). Prostate cancer risk in US blacks and whites with a family history of cancer. *Int. J. Cancer, 60*, 361.

Imran, A., Waseem, A. W., & Kishwar, S., (2011). Cancer scenario in India with future perspectives. *Cancer Ther., 8*, 56–70.

Jiang, Z., Wu, C. L., Woda, B. A., Dresser, K., Xu, J., Fanger, G. R., & Yang, X. J., (2002). P504S/alpha-methylacyl-CoA racemase: A useful marker for diagnosis of small foci of prostatic carcinoma on needle biopsy. *Amer. J. Surg. Pathol., 26*, 1169–1174.

Jiang, Z., Wu, C. L., Woda, B. A., Iczkowski, K. A., Chu, P. G., Tretiakova, M. S., Young, R. H., Weiss, L. M., Blute, Jr. R. D., Brendler, C. B., & Krausz, T., (2004). Alpha-methylacyl-CoA racemase: A multi-institutional study of a new prostate cancer marker. *Histopathol., 45*, 218–225.

Kirby, R. S., Fitzpatrick, J. M., & Irani, J., (2009). Prostate cancer diagnosis in the new millennium: Strengths and weaknesses of prostate-specific antigen and the discovery and clinical evaluation of prostate cancer gene 3 (PCA3). *Br. J. Urol. Int., 103*, 441–445.

Kuefer, R., Varambally, S., Zhou, M., Lucas, P. C., Loeffler, M., Wolter, H., Mattfeldt, T., Hautmann, R. E., Gschwend, J. E., Barrette, T. R., & Dunn, R. L., (2002). Alpha-methylacyl-CoA racemase: Expression levels of this novel cancer bio marker depend on tumor differentiation. *Amer. J. Pathol., 161*, 841–848.

Lallet-Daher, H., Roudbaraki, M., Bavencoffe, A., Mariot, P., Gackiere, F., Bidaux, G., Urbain, R., Gosset, P., Delcourt, P., Fleurisse, L., & Slomianny, C., (2009). Intermediate-conductance Ca2+ activated K+ channels (IKcal) regulate human prostate cancer cell proliferation through a close control of calcium entry. *Oncogene, 28*, 1792–1806.

Lee, W. H., Morton, R. A., Epstein, J. I., Brooks, J. D., Campbell, P. A., Bova, G. S., Hsieh, W. S., Isaacs, W. B., & Nelson, W. G., (1994). Cytidine methylation of regulatory sequences near the pi-class glutathione S-transferase gene accompanies human prostatic carcinogenesis. *Proc. Natl. Acad. Sci. U.S.A., 91*, 11733–11771.

Li, Y., & Sarkar, F. H., (2002). Inhibition of nuclear factor kappa B activation in PC3 cells by genistein is mediated via Akt signaling pathway. *Clin. Cancer Res., 8*, 2369–2377.

Liang, J. Y., Liu, Y. Y., Zou, J., Franklin, R. B., Costello, L. C., & Feng, P., (1999). Inhibitory effect of zinc on human prostatic carcinoma cell growth. *Prostate, 40*, 200–207.

Liao, J., Schneider, A., Datta, N. S., & McCauley, L. K., (2006). Extracellular calcium as a candidate mediator of prostate cancer skeletal metastasis. *Cancer Res., 77*, 9065–9073.

Markowski, M. C., Bowen, C., & Gelmann, E. P., (2008). Inflammatory cytokines induce phosphorylation and ubiquitination of prostate suppressor protein NKX3.1. *Cancer Res., 68*, 6896–6901.

Okamoto, T., Schlegel, A., Scherer, P. E., & Lisanti, M. P., (1998). Caveolins, a family of scaffolding proteins for organizing "preassembled signaling complexes" at the plasma membrane. *J. Biol. Chem., 273*, 5419–5422.

Parkin, D. M., Bray, F., Ferlay, J., & Pisani, P., (2005). Global cancer statistics, 2002. *CA Cancer J. Clin., 55*, 74–108.

Rubin, M. A., Zhou, M., Dhanasekaran, S. M., Varambally, S., Barrette, T. R., Sanda, M. G., Pienta, K. J., Ghosh, D., & Chinnaiyan, A. M., (2002). Alpha-methylacyl coenzyme A racemase as a tissue biomarker for prostate cancer. *J. Amer. Med. Assoc., 287*, 1662–1670.

Salami, S. S., Schmidt, F., Laxman, B., Regan, M. M., Rickman, D. S., Scherr, D., Bueti, G., Siddiqui, J., Tomlins, S. A., Wei, J. T., & Chinnaiyan, A. M., (2011). Combining urinary detection of TMPRSS2: ERG and PCA3 with serum PSA to predict diagnosis of prostate cancer. *Urol. Oncol., 31*, 566–571.

Scholzen, T., & Gerdes, J., (2000). The Ki-67 protein: From the known and the unknown. *J. Cell. Physiol., 182*, 311–322.

Schwartz, G. G., (2009). Is serum calcium a bio marker of fatal prostate cancer? *Futur. Oncol., 5*, 577–580.

Shroeder, F. H., (1999). Endocrine treatment of prostate cancer—recent developments and the future. Part 1: Maximal androgen blockade, early vs. delayed endocrine treatment and side-effects. *Br. J. Urol. Int., 83*, 161–170.

Simon, J. P., & Aunis, D., (1989). Biochemistry of the chromogranin A protein family. *J. Biochem., 262*, 1–13.

Sinha, R., & El-Bayoumy, K., (2004). Apoptosis is a critical cellular event in cancer chemoprevention and chemotherapy by selenium compounds. *Curr. Cancer Drug Targets, 4*, 13–28.

Sporn, M. B., & Suh, N., (2002). Chemoprevention: An essential approach to controlling cancer. *Nat. Rev. Cancer, 2*, 537–543.

Sporn, M. B., (1976). Approaches to prevention of epithelial cancer during the preneoplastic period. *Cancer Res., 36*, 2699–2702.

Sreekumar, A., Poisson, L. M., Rajendiran, T. M., Khan, A. P., Cao, Q., Yu, J., Laxman, B., Mehra, R., Lonigro, R. J., Li, Y., & Nyati, M. K., (2009). Metabolomic profiles delineate potential role for sarcosine in prostate cancer progression. *Nature, 457*, 799–800.

Tahir, S. A., Yang, G., Goltsov, A. A., Watanabe, M., Tabata, K. I., Addai, J., Kadmon, D., & Thompson, T. C., (2008). Tumor cell-secreted caveolin-1 has pro-angiogenic activities in prostate cancer. *Cancer Res., 68*, 731–739.

Taper, H. S., Jamison, J. M., Gilloteaux, J., Gwin, C. A., Gordon, T., & Summers, J. L., (2001). *In vivo* reactivation of DNases in implanted human prostate tumors after administration of a vitamin C/K (3) combination. *J. Histochem. Cytochem., 49*, 109–120.

Teng, D. H. F., Hu, R., Lin, H., Davis, T., Iliev, D., Frye, C., Swedlund, B., Hansen, K. L., Vinson, V. L., Gumpper, K. L., & Ellis, L., (1997). MMAC1/PTEN mutations in primary tumor specimens and tumor cell lines. *Cancer Res., 57*, 5221–5225.

Tirado, O. M., Maccarthy, C. M., Fatima, N., Villar, J., Mateo-Lozano, S., & Notario, A. V., (2010). Caveolin-1 promotes resistance to chemotherapy-induced apoptosis in Ewing's sarcoma cells by modulating PKC alpha phosphorylation. *Int. J. Cancer, 126*(2), 426–436.

Vainio, H., (1999). Chemoprevention of cancer: a controversial and instructive story. *Br. Med. Bull., 55*, 593–599.

Vayalil, P. K., Mittal, A., & Katiyar, S. K., (2004). Proanthocyanidins from grape seeds inhibit expression of matrix metalloproteinases in human prostate carcinoma cells, which is associated with the inhibit ion of activation of MAPK and NF kappa B. *Carcinogenesis, 25*, 987–995.

Velonas, M. V., Woo, H. H., Dos, R. G. C., & Assinder, J. S., (2013). Current status of biomarkers for prostate cancer. *Int. J. Mol. Sci., 14*, 11034–11060.

Wang, Z., Tseng, C. P., Pong, R. C., Chen, H., McConnell, J. D., Navone, N., & Hsieh, J. T., (2002). The mechanism of growth-inhibitory effect of DOC-2/ DAB2 in prostate cancer. Characterization of a novel GTPase-activating protein associated with N-terminal domain of DOC-2/DA B2. *J. Biol. Chem., 277*, 12622–12631.

Weisburger, J. H., (1999). Mechanisms of action of antioxidants as exemplified in vegetables, tomatoes, and tea. *Food Chem. Toxicol., 37*, 943–948.

Young, L., Percy, C. L., Asire, A. J., Berg, J. W., Cusano, M. M., Gloeckler, L. A., Horm, J. W., Lourie, Jr. W. I., Pollack, E. S., & Shambaugh, E. M., (1981). Cancer incidence and mortality in the United States-1973–1977. *J. Natl. Cancer Inst. Monogr., 57*, 1001–1081.

Zha, S., Ferdinandusse, S., Denis, S., Wanders, R. J., Ewing, C. M., Luo, J., De Marzo, A. M., & Isaacs, W. B., (2003). Alpha-methylacyl-CoA racemase as an androgen-independent growth modifier in prostate cancer. *Cancer Res., 63*, 7365–7376.

Zhi, G., Sergei, K., Martin, L. F., Maria, P. D., & Tanya, P. T., (2010). ATM Activation by oxidative stress. *Sci., 330*, 517.

CHAPTER 16

Overview of Tuberculosis Biomarkers: An Update

ABHILASHA TRIPATHI,[1] SURYA KANT,[1]
PREM PRAKASH KUSHWAHA,[2] and SHASHANK KUMAR[2]

[1]*Department of Respiratory and Medicine,
King George's Medical University, Lucknow, India.*

[2]*School of Basic and Applied Sciences, Department of Biochemistry,
Central University of Punjab, Bathinda, Punjab–151001, India,
Tel.: +91 9335647413, E-mail: shashankbiochemau@gmail.com*

ABSTRACT

There are millions of cases of tuberculosis (TB) get undiagnosed annually. New tactics and innovative diagnostic tools to control TB are urgently needed throughout the world. Exact diagnosis and early treatment of TB and latent TB infection are dynamic to prevent and control the TB. The absence of precise biomarkers hampers these efforts. In particular, new diagnosis tools and biomarkers are compulsory to estimate both pathogen and host key elements of the response to TB infection. Multiple-marker bio-signature or biomarker based tests, ideally achieved in urine or blood sample. Biomarker based active TB recognition might meet target product profiles anticipated by the WHO for point-of-care testing. Current efforts to recognize the TB diagnosis have exposed the crucial need for biomarker-based protocol that will allow more affordable, efficient, and accessible diagnosis for those in need. In the present chapter, we summarizes the different types of biomarkers and biosignatures such as serum/plasma biomarker [MCP-3 (monocyte chemoattractant protein-3), IL-4, IFN-Υ, EGF, fractalkine, IP-10], antigen stimulated biomarker (CFP10, ESAT6, TB7.7), micro-RNA based biomarker (miR-378, miR-29c, miR-483, miR-193a-5p, miR-148a, miR-192, miR-146a, let-7e, miR-178, miR-365), etc.

16.1   INTRODUCTION

Pathogenic bacteria release pathogenic substances which cause severe infections and disease. Mycobacterium tuberculosis (MTB) bacteria cause Tuberculosis (TB) which is a treatable disease. It has two stages namely active TB and latent disease (Field, 2001). Active TB in larynx or lungs may infect others through the sneezing or coughing. Inadequate treatment is the major cause of relapse or drug resistance. HIV infection is another stimulated cause of the revival of TB. According to a report, a number of patient's caretakers died due to the infection from the suffering person (Dooley and Tapper, 1997). These caretakers were suffered from poor immune system with some ailments such as HIV infection or AIDS. Latent TB infection comprises less or no symptoms, no feeling of sick, possess positive TB skin and TB blood test, normal chest x-ray and negative sputum smear. A person with TB disease comprises feeling of sick, weight loss, pain in the chest, chills, fever, sweating at night, abnormal chest x-ray and positive sputum smear or culture. TB infection starts with mycobacteria inhalation into the lung alveoli. Some of the infected person having strong immunity can clear the infection by utilizing the innate or adaptive immune mechanisms. Based on this mechanism, two tests such as tuberculin skin test (TST) and IFN-g release assay (IGRA)) are available to detect the active TB. The person having poor immunity may suffer with latent TB or active TB. People who have the ability to clear the infection may re-infect with TB depends upon the host immunity power. For the treatment of TB, BCG vaccine is available with their high effectiveness. A large number of drugs have been reported for the drug-sensitive TB, MDR-TB, and XDR-TB (Table 16.1).

16.2   IMMUNOLOGICAL CHARACTERISTICS

MTB infection requires three to eight-week starting immunity development. This can be identified by late hypersensitivity reaction which gives a positive reaction for the TST. Infected individuals comprise latent tuberculosis (TB) but later only 10% active TB develops. Immunological protection against TB includes both innate and adaptive responses. T helper 1 (Th1) cells primarily direct the response processes against TB (Ottenhoff, 2012a; Ottenhoff and Kaufmann, 2012).

TABLE 16.1 Different Generation of Drugs to Treat the Active TB Infection

First-Line Treatment of TB for Drug-Sensitive TB	Multidrug-Resistant Tuberculosis (MDR TB) and Second-Line Treatments	Extensively Drug-Resistant Tuberculosis (XDR TB)-Options for Treatment	New Candidate TB Drugs Under Development
Isoniazid (1952)	*Isoniazid*	*Isoniazid*	SQ-109
Rifampin (1966)	*Rifampin*	*Rifampin*	Meropenem (*Injection*)
Ethambutol (1961)	Thioamides: Ethionamide and Prothionamide	*Fluoroquinolones*	Imidazopyridine amide: Q203
Pyrazinamide (1952)	Diarylquinoline: Bedaquiline (TMC-207)	*Injectable Second Line Drugs (Bacteria are resistant to at least one): Kanamycin, Amikacin, and Capreomycin*	Benzothiazinones: PBTZ169 and BTZ043
	Cyclic peptides: Capreomycin (*injection*)	Newly introduced drug: Bedaquiline and Delamanid	Oxazolidinones: Sutezolid and Linezolid
	Nitroimidazole: Delamanid (OPC-67683)	Possible effective drugs: Ethambutol, Pyrazinamide, Thioamides, Cycloserine, Para-aminosalicylic acid (PAS), Streptomycin, Clofazimine	Rifapentine
	Ethambutol		
	Pyrazinamide (PAZ)		Macrolides
	Para-aminosolicyclic acid (PAS)		Nitroimidazoles: PA-824
	Fluoroquinolones: Moxifloxacin and Levofloxacin		
	Cycloserine		
	Aminoglycosides: Kanamycin, Amikacin, and Streptomycin (*Injection*)		

TABLE 16.1 *(Continued)*

Mechanisms of Action of Current TB Drugs

Drug	Mechanism of Action
Thioamides	Inhibit cell wall synthesis
Nitroimidazoles	Inhibit mycolic acid synthesis
Ethambutol	Inhibit cell wall synthesis
Cycloserine	Inhibit cell wall synthesis
Pyrazinamide	Exact target is unclear, Disrupt plasma membrane, Disrupts energy metabolism
Diarylquinoline	Inhibit ATP synthesis
PAS	Inhibit synthesis of DNA precursors
Fluoroquinolones	Inhibit DNA gyrase
Cyclic peptides	Inhibit protein synthesis
Aminoglycosides	Inhibit protein synthesis

Mechanisms of Action of TB Drugs Under Development

Drug	Mechanism of Action
Nitroimidazoles	Inhibit mycolic acid and other targets
SQ-109	Inhibit cell wall synthesis
Meropenem	Inhibit peptidoglycan synthesis
Benzothiazinones	Inhibit cell wall synthesis
Imidazopyridine amide	Inhibits cytochrome oxidase
Rifamycins (Rifapentine)	Inhibits RNA synthesis
Oxazolidinones (Linezolid, Sutezolid)	Inhibit protein synthesis
Macrolides	Inhibit protein synthesis

Drug names written in italics are resistant to bacteria in a particular generation.

16.2.1 SERUM/PLASMA BIOMARKER

Serum/plasma biomarker measurement explores various kinds of infections in particular patients. Cytokine levels in serum have been reported for latent and active TB. MCP-3 (monocyte chemoattractant protein-3), IL-4, IFN-Υ, EGF, fractalkine, and IP-10 have been shown significantly to TB patients and healthy individuals (Mihret et al., 2013). Another study demonstrated that proinflammatory cytokines IL-6 and IP-10 in serum were down expressed in latent infected individuals compare to the TB patients (DjobaSiawaya et al., 2009a). In the same study, they also demonstrated that MCP-1 levels were increased in latent infected individuals compared to the TB patients (DjobaSiawaya et al., 2009a). Granulysin levels in plasma also showed significant treatment response (Sahiratmadja et al., 2007). The like soluble CD14 (sCD14), toll-like receptor 4 pathway and myeloid differentiation-2 (MD-2) levels in plasma also discriminate the latent infected individuals and TB patients (Feruglio et al., 2013). Cytokine expression pattern can also detect month-2 culture conversion and able to show the treatment response and relapse (Walzl et al., 2011).

16.2.2 ANTIGEN STIMULATED BIOMARKER

IFN-Υ release assays (IGRAs) establishes the release of IFN-Υ after stimulation of peripheral blood mononuclear cells (PBMCs) in answer to Mtb-definite antigens such as CFP10, ESAT6, and TB7.7 (Lalvani et al., 2001). This is the best assay to detect the infection of TB instead of less precise TST (Janssens, 2007). IGRAs test fails to discriminate between active and latent. In a study, increased expression of IFN-Υ reported for the latent infection just formerly to TB diagnosis (Diel et al., 2008; Higuchi et al., 2008). Various cytokines levels stimulated by pathogen proteins can discriminate between active TB and latent. Detection of the soluble CD40 ligand (sCD40L), epidermal growth factor (EGF), macrophage inflammatory protein-1b (MIP-1b), transforming growth factor-α, vascular endothelial growth factor (VEGF) and IL-1a can also differentiate between active and latent TB (Chegou et al., 2009). Biomarkers such as IL-1Ra, IFN-a2, IP-10, sCD40L, and VEGF can also discriminate between active and latent TB in HIV-positive and negative children (Chegou et al., 2013).

16.3   M. TUBERCULOSIS DERIVED BIOMARKERS

Biomarkers for the infectious disease generally comprise host/pathogen-derived (McNerney et al., 2012). Before detecting any targeted antigen, pathogen-derived moiety should be reached in samples such as plasma, urine, and sputum in sufficient amount (Bekmurzayeva et al., 2013). For the detection of TB, several samples can be used such as blood, sputum, saliva, urine, pleural fluid, and cerebrospinal fluid (CSF). Urine test for the TB detection is much easier due to the handling and collection as well as less variable comparison to sputum. The urine test can also detect the TB in HIV co-infected patients (WHO, 2009). Lipoarabinomannan (LAM) is the major antigen is being tested for TB identification. LAM required for the formation of an outer layer of the cell wall in the Mycobacterium genus. It originates from degrading cells or metabolically active cells which are later cleared by kidney cells (Hunter et al., 1986). Detection of LAM can be achieved by a sandwich capture ELISA method in sputum or urine sample (Mutetwa et al., 2009). Recent advancement in therapeutic approaches defines a broad spectrum for the discovery of biomarker-based on the proteomic approaches by using the sample of TB patients. Four M. tuberculosis proteins namely homoserine O-acetyl transferase, ornithine carbamoyltransferase, molybdopterin biosynthesis protein, and 3-phosphoadenosine-5-phosphosulfate reductase were identified in different studies by using the urine sample (Napolitano et al., 2008). Recently a study reported that gene coding MoeX protein can be used as a diagnostic biomarker for pulmonary TB (Pollock et al., 2013). List of proteins involved in pathogenesis is summarized in Table 16.2.

16.4   TB TREATMENT RESPONSE BIOMARKER

16.4.1   *PATHOGEN BASED BIOMARKER*

To monitor the sputum burden of Mtb, quantification of microbes is a suitable method. According to researchers, up to 90% of Mtb in the sputum of treatment-naïve patients fails to grow on the agar plate. Quantification of the bacilli demonstrates the drug-tolerant populations that create a delay in treatment (Chengalroyen et al., 2016). Destruction of bacterial mRNA takes few seconds which suggests that mRNA signals finding may evidence for the occurrence of viable Mtb. The molecular bacterial load (MBL) assay reveals the quantification of Mtb 16S rRNA in a precise manner. To detect the bactericidal activity, MBL also established as equivalence to culture on solid agar (Honeyborne et al., 2011). In a study, outcomes denoted that triacylglycerol lipid bodies gathering suggests a non-replicating, drug-tolerant bacterial phenotype which may be the major reason of the failure of treatment and relapse (Sloan et al., 2015). In another study, staining of sputum acid-fast bacillus lipid body was increased initial in treatment in patients who ensued to failure and relapse than who were cured (Sloan et al., 2015).

TABLE 16.2 Summary of Principal Protein Antigens Evaluated for *Mycobacterium Tuberculosis* Direct Diagnosis

Protein Information	Function	References
Alanine proline-rich secreted protein APA	Provide bacterial attachment to host cell	Tucci et al., 2014
ESAT-6 (Early secretory antigen target 6)	Produces increased levels of IFN-Υ from memory effector cells.	
Secreted antigen 85-A FBPA (fibronectin-binding protein A), Secreted antigen 85-B FBPB (fibronectin-binding protein B)	Take parts in cell wall mycoloylation. Makes mycobacteria fibronectin attachment feasible.	
Malate synthase G	Involved in glyoxylate bypass, substitute to the TCA cycle	
Chaperonin 2	Avoids misfolding and indorse appropriate folding	
Heat shock protein	Provide long-term viability during latent and asymptomatic infections and in replication during early infection.	
Possible molybdopterin biosynthesis protein	Play role in molybdopterin cofactor synthesis.	
Immunogenic protein MPT64	Function is unknown but specific for *M. tuberculosis.*	
Periplasmic phosphate-binding lipoprotein PSTS1	Provide active transport of inorganic phosphate throughout the membrane	
Universal stress protein family protein TB31.7.	Controls mycobacterial growth	

16.4.2   HOST-BASED BIOMARKER

Host biomarker used for the measurement of the treatment response. Study of gene expression on large scale in TB patients, it has been reported that type I and type II IFN signaling expression get normalized after treatment (Ottenhoff et al., 2012). In another study, transcriptomic expression of relapsing versus non-relapsing patients demonstrated that increased cytosolic response predicts relapse even after effective TB treatment (Cliff et al., 2015). Host biomarkers such as chemokines and cytokines (IFN-γ, TNF-α, IP-10, and IL-10) can predict treatment response. Increased TNF-α levels indicated in the way of positive response against treatment (Ameglio et al., 2005) while increased IL- 10 levels correlated with shorter survival (Wang et al., 2012). Biomarkers associated with apoptosis also demonstrated that increased monocyte chemotactic protein (MCP)-1 and serum decoy receptor 3 (DcR3) were autonomously related with poorer six-month survival in TB patients (Shu et al., 2013).

16.5   BIOMARKERS FOR TB RELAPSE AFTER TREATMENT

TB relapse comprises the recrudescence of disease after the apparently effective treatment. After treatment, this can occur for up to two years. Nowadays, failure of sputum culture conversion within two months after the start of anti-TB treatment act as a prognostic indicator for high risk of TB relapse subsequent to the treatment of pulmonary TB (Wallis et al., 2009). Both positive prognostic value and sensitivity are insufficient to assure the clinically valuable measure of risk. Some of the other biomarkers comprise numbers of colony-forming units in Mtb culture and pretreatment sputum bacillary load (Wallis et al., 2009). Parenchymal contribution and cavitation extent, time to detection by the culture at early diagnosis and sputum smear grade were recorded which positively correlated with culture conversion at two months (Hesseling et al., 2010). Some of the studies also demonstrated the immunological biomarkers for relapse. IL-17 levels and serum IP-10 were reported for the positive association with sputum smear/culture results two months after treatment. Serum IP-10 and IL-17 both represent predicted risk of relapse and mortality respectively (Chen et al., 2011). Comparative to other prognostic biomarkers, identification, and assessment of relapse biomarker are predominantly challenging because it requires long-term follow-up of patients inadequate number.

16.6 PROTEOMIC, TRANSCRIPTOMIC, AND METABOLOMIC BIOMARKERS

System biology approaches explored different diagnostic biomarkers for vigorous ailments in recent years. Microarray analysis of blood cells defines a transcriptomic expression pattern that keeps remarkable importance for disease diagnosis. Berry et al. (2010) demonstrated 393-gene in active TB patients and 86-transcript that positively correlates with different bacterial and autoimmune diseases. In another study, 927 genes in active TB versus LTBI/noninfected control positively supported the diagnostic assay (Maertzdorf et al., 2011). PBMCs genome-wide gene-expression analysis was also reported as prognostic biomarkers for active TB (Ottenhoff et al., 2012). Bioinformatics accessories established the platform for precisely distinguished active TB from LTBI and controls (Maertzdorf et al., 2011). Recent advancement in proteomic approach through biological tools and techniques such as matrix-assisted laser ionization/desorption time-of-flight mass spectrometry (MS) and surface- enhanced laser ionization/ desorption time-of-flight MS supports in the identification of active TB biomarkers form the blood or body fluids (Fu et al., 2012). In a study, researchers demonstrated the biomarker levels for extrapulmonary TB, smear-negative TB, and controls (Fu et al., 2012). Metabolic changes recorded during the TB infection originated by mycobacterium, deliver a new policy for crucial diagnostic biomarkers. Several studies already reported on metabolic changes in the animals for the identification of the ailments (Shin et al., 2011).

16.7 MICRORNA-BASED BIOMARKER

RNA sequencing comprises host microRNA (miRNA) identification which can act as a biomarker for active TB and pulmonary TB. Around 15 miRNAs present in a serum sample can distinguish the latent TB and pulmonary TB with the precision of about 82% (Miotto et al., 2013). Another study reported that six miRNA in serum sample shown specificity and sensitivity of around 92% and 95% respectively to diagnose TB (Zhang et al., 2013). Some of the miRNAs which act as a biomarker for the TB patients are the miR-378, miR-29c, miR-483, miR-193a-5p, miR-148a, miR-192, miR-146a, let-7e, miR-178, and miR-365 (Miotto et al., 2013; Zhang et al., 2013).

16.8   CHALLENGES IN TB BIOMARKER DEVELOPMENT

Identification and validation of the biomarker which could be rendered for the sensitive and precise diagnostic test for TB is a challenging task for researchers. Knowledge gaps in the fields of understanding of the ailment and host-pathogen interaction require the urgent need of identification of biomarkers to diagnose the disease. Specific molecular biomarkers identification would help in the *in vitro* diagnostic test development for the *M. tuberculosis* infection. These biomarkers should be inexpensive, appropriate, rapid, and sensitive to be used in test centers with low cost and structure.

KEYWORDS

- cerebrospinal fluid
- decoy receptor 3
- epidermal growth factor
- fibronectin-binding protein A
- lipoarabinomannan
- molecular bacterial load

REFERENCES

Ameglio, F., Casarini, M., Capoluongo, E., Mattia, P., Puglisi, G., & Giosue, S., (2005). Post- treatment changes of six cytokines in active pulmonary tuberculosis: Differences between patients with stable or increased fibrosis. *Int. J. Tuberc. Lung Dis.*, *9*(1), 98–104.

Bekmurzayeva, A., Sypabekova, M., & Kanayeva, D., (2013). Tuberculosis diagnosis using immunodominant, secreted antigens of mycobacterium tuberculosis. *Tuberculosis*, *93*(4), 381–388.

Berry, M. P., Graham, C. M., McNab, F. W., Xu, Z., Bloch, S. A., Oni, T., Wilkinson, K. A., Banchereau, R., Skinner, J., Wilkinson, R. J., & Quinn, C., (2010). An interferon-inducible neutrophil-driven blood transcriptional signature in human tuberculosis. *Nature*, *466*(7309), 973–977.

Chegou, N. N., Black, G. F., Kidd, M., Van, H. P. D., & Walzl, G., (2009). Host markers in QuantiFERON supernatants differentiate active TB from latent TB infection: Preliminary report. *BMC Pulm. Med.*, *9*(1), 1–12.

Chegou, N. N., Detjen, A. K., Thiart, L., Walters, E., Mandalakas, A. M., Hesseling, A. C., & Walzl, G., (2013). Utility of host markers detected in quantiferon supernatants for the diagnosis of tuberculosis in children in a high-burden setting. *PloS One*, *8*(5), 1–11.

Chen, Y. C., Chin, C. H., Liu, S. F., Wu, C. C., Tsen, C. C., Wang, Y. H., Chao, T. Y., Lie, C. H., Chen, C. J., Wang, C. C., & Lin, M. C., (2011). Prognostic values of serum IP-10 and IL-17 in patients with pulmonary tuberculosis. *Dis. Markers*, *31*(2), 101–110.

Chengalroyen, M. D., Beukes, G. M., Gordhan, B. G., Streicher, E. M., Churchyard, G., Hafner, R., Warren, R., Otwombe, K., Martinson, N., & Kana, B. D., (2016). Detection and quantification of differentially culturable tubercle bacteria in sputum from patients with tuberculosis. *Am. J. Respir. Crit. Care Med.*, *194*(12), 1532–1540.

Cliff, J. M., Cho, J. E., Lee, J. S., Ronacher, K., King, E. C., Van, H. P., Walzl, G., & Dockrell, H. M., (2015). Excessive cytolytic responses predict tuberculosis relapse after apparently successful treatment. *J. Infect. Dis.*, *213*(3), 485–495.

Diel, R., Loddenkemper, R., Meywald-Walter, K., Niemann, S., & Nienhaus, A., (2008). Predictive value of a whole blood IFN-γ assay for the development of active tuberculosis disease after recent infection with Mycobacterium tuberculosis. *Am. J. Respir. Crit. Care Med.*, *177*(10), 1164–1170.

DjobaSiawaya, J. F., Beyers, N., Van, H. P., & Walzl, G., (2009a). Differential cytokine secretion and early treatment response in patients with pulmonary tuberculosis. *Clin. Exp. Immunol.*, *156*(1), 69–77.

Dooley, S. W., & Tapper, M., (1997). Epidemiology of nosocomial tuberculosis. In: Wenzel, R. P., (ed.), *Prevention and Control of Nosocomial Infections* (3rd edn., pp. 357–394). Baltimore: Williams & Wilkins.

Feruglio, S. L., Trøseid, M., Damås, J. K., Kvale, D., & Dyrhol-Riise, A. M., (2013). Soluble markers of the toll-like receptor 4 pathway differentiate between active and latent tuberculosis and are associated with treatment responses. *PloS One*, *8*(7), 1–8.

Field, M. J., (2001). *Tuberculosis in the Workplace*. National Academies Press.

Fu, Y. R., Yi, Z. J., Guan, S. Z., Zhang, S. Y., & Li, M., (2012). Proteomic analysis of sputum in patients with active pulmonary tuberculosis. *Clin. Microbiol. Infect.*, *18*(12), 1241–1247.

Hesseling, A. C., Walzl, G., Enarson, D. A., Carroll, N. M., Duncan, K., Lukey, P. T., Lombard, C., Donald, P. R., Lawrence, K. A., Gie, R. P., & Van, H. P. D., (2010). Baseline sputum time to detection predicts month two culture conversion and relapse in non-HIV-infected patients. *Int. J. Tuberc. Lung Dis.*, *14*(5), 560–570.

Higuchi, K., Harada, N., Fukazawa, K., & Mori, T., (2008). Relationship between whole-blood interferon-g responses and the risk of active tuberculosis. *Tuberculosis (Edinb)*, *88*, 244–248.

Honeyborne, I., McHugh, T. D., Phillips, P. P., Bannoo, S., Bateson, A., Carroll, N., Perrin, F. M., Ronacher, K., Wright, L., Van, H. P. D., & Walzl, G., (2011). Molecular bacterial load assay, a culture-free biomarker for rapid and accurate quantification of sputum mycobacterium tuberculosis bacillary load during treatment. *J. Clin. Microbiol.*, *49*(11), 3905–3911.

Hunter, S. W., Gaylord, H., & Brennan, P. J., (1986). Structure and antigenicity of the phosphorylated lipopolysaccharide antigens from the leprosy and tubercle bacilli. *J. Biol. Chem.*, *261*(26), 12345–12351.

Janssens, J. P., (2007). Interferon-g release assay tests to rule out active tuberculosis. *Eur. Respir. J.*, *30*, 183–184.

Lalvani, A., Pathan, A. A., Durkan, H., Wilkinson, K. A., Whelan, A., Deeks, J. J., Reece, W. H., Latif, M., Pasvol, G., & Hill, A. V., (2001). Enhanced contact tracing and spatial tracking of Mycobacterium tuberculosis infection by enumeration of antigen-specific T-cells. *Lancet*, *357*(9273), 2017–2021.

Maertzdorf, J., Repsilber, D., Parida, S. K., Stanley, K., Roberts, T., Black, G., Walzl, G., & Kaufmann, S. H., (2011). Human gene expression profiles of susceptibility and resistance in tuberculosis. *Genes Immun.*, *12*(1), 15–22.

McNerney, R., Maeurer, M., Abubakar, I., Marais, B., Mchugh, T. D., Ford, N., Weyer, K., Lawn, S., Grobusch, M. P., Memish, Z., & Squire, S. B., (2012). Tuberculosis diagnostics and biomarkers: Needs, challenges, recent advances, and opportunities. *J. Infect. Dis.*, *205*(2), S147–S158.

Mihret, A., Bekele, Y., Bobosha, K., Kidd, M., Aseffa, A., Howe, R., & Walzl, G., (2013). Plasma cytokines and chemokines differentiate between active disease and non-active tuberculosis infection. *J. Infect.*, *66*(4), 357–365.

Miotto, P., Mwangoka, G., Valente, I. C., Norbis, L., Sotgiu, G., Bosu, R., Ambrosi, A., Codecasa, L. R., Goletti, D., Matteelli, A., & Ntinginya, E. N., (2013). miRNA signatures in sera of patients with active pulmonary tuberculosis. *PLoS One*, *8*(11), 1–15.

Mutetwa, R., Boehme, C., Dimairo, M., Bandason, T., Munyati, S. S., Mangwanya, D., Mungofa, S., Butterworth, A. E., Mason, P. R., & Corbett, E. L., (2009). Diagnostic accuracy of commercial urinary lipoarabinomannan detection in African tuberculosis suspects and patients. *Int. J. Tuberc. Lung Dis.*, *13*(10), 1253–1259.

Napolitano, D. R., Pollock, N., Kashino, S. S., Rodrigues, V., & Campos-Neto, A., (2008). Identification of mycobacterium tuberculosis ornithine carboamyltransferase in urine as a possible molecular marker of active pulmonary tuberculosis. *Clin. Vaccine Immunol.*, *15*(4), 638–643.

Ottenhoff, T. H. M., (2012a). The knowns and unknowns of the immune pathogenesis of tuberculosis. *Int. J. Tuberc. Lung Dis.*, *16*(11), 1424–1432.

Ottenhoff, T. H., Dass, R. H., Yang, N., Zhang, M. M., Wong, H. E., Sahiratmadja, E., Khor, C. C., Alisjahbana, B., Van, C. R., Marzuki, S., & Seielstad, M., (2012). Genome-wide expression profiling identifies type 1 interferon response pathways in active tuberculosis. *PloS One*, *7*(9), 1–12.

Pollock, N. R., Macovei, L., Kanunfre, K., Dhiman, R., Restrepo, B., Zarate, I., Pino, P. A., Mora-Guzman, F., Fujiwara, R. T., Michel, G., & Kashino, S. S., (2013). Validation of Mycobacterium tuberculosis Rv1681 protein as a diagnostic marker of active pulmonary tuberculosis. *J. Clin. Microbiol.*, *51*(5), 1367–73.

Sahiratmadja, E., Alisjahbana, B., Buccheri, S., Di Liberto, D., De Boer, T., Adnan, I., Van, C. R., Klein, M. R., Van, M. K. E., Nelwan, R. H. H., & Van, D. V. E., (2007). Plasma granulysin levels and cellular interferon-γ production correlate with curative host responses in tuberculosis, while plasma interferon-γ levels correlate with tuberculosis disease activity in adults. *Tuberculosis*, *87*(4), 312–321.

Shin, J. H., Yang, J. Y., Jeon, B. Y., Yoon, Y. J., Cho, S. N., Kang, Y. H., Ryu, D. H., & Hwang, G. S., (2011). 1H NMR-based metabolomic profiling in mice infected with mycobacterium tuberculosis. *J. Proteome Res.*, *10*(5), 2238–2247.

Shu, C. C., Wu, M. F., Hsu, C. L., Huang, C. T., Wang, J. Y., Hsieh, S. L., Yu, C. J., Lee, L. N., & Yang, P. C., (2013). Apoptosis-associated biomarkers in tuberculosis: Promising for diagnosis and prognosis prediction. *BMC Infect. Dis.*, *13*(1), 1–7.

Sloan, D. J., Mwandumba, H. C., Garton, N. J., Khoo, S. H., Butterworth, A. E., Allain, T. J., Heyderman, R. S., Corbett, E. L., Barer, M. R., & Davies, G. R., (2015). Pharmacodynamic modeling of bacillary elimination rates and detection of bacterial lipid bodies in sputum to predict and understand outcomes in treatment of pulmonary tuberculosis. *Clin. Infect. Dis.*, *61*(1), 1–8.

Tucci, P., González-Sapienza, G., & Marin, M., (2014). Pathogen-derived biomarkers for active tuberculosis diagnosis. *Front. Microbiol.*, *5*, 1–6.

Wallis, R. S., Doherty, T. M., Onyebujoh, P., Vahedi, M., Laang, H., Olesen, O., Parida, S., & Zumla, A., (2009). Biomarkers for tuberculosis disease activity, cure, and relapse. *Lancet Infect. Dis.*, *9*(3), 162–172.

Walzl, G., Ronacher, K., Hanekom, W., Scriba, T. J., & Zumla, A., (2011). Immunological biomarkers of tuberculosis. *Nat. Rev. Immunol.*, *11*(5), 343–54.

Wang, J. Y., Chang, H. C., Liu, J. L., Shu, C. C., Lee, C. H., Wang, J. T., & Lee, L. N., (2012). Expression of toll-like receptor 2 and plasma level of interleukin-10 are associated with outcome in tuberculosis. *Eur. J. Clin. Microbiol. Infect. Dis.*, *31*(9), 2327–2333.

World Health Organization (WHO), (2009). *Pathways to Better Diagnostics for Tuberculosis: A Blue Print for the Development of Tb Diagnostics by the New Diagnostics Working Group of the Stop Tb Partnership*. Geneva: World Health Organization.

Zhang, X., Guo, J., Fan, S., Li, Y., Wei, L., Yang, X., Jiang, T., Chen, Z., Wang, C., Liu, J., & Ping, Z., (2013). Screening and identification of six serum microRNAs as novel potential combination biomarkers for pulmonary tuberculosis diagnosis. *PloS One*, *8*(12), 1–11.

Index

Printed and bound by CPI Group (UK) Ltd, Croydon, CR0 4YY

23/10/2024

01777702-0002